HOME-ORIENTED INFORMATICS AND TELEMATICS

IFIP – The International Federation for Information Processing

IFIP was founded in 1960 under the auspices of UNESCO, following the First World Computer Congress held in Paris the previous year. An umbrella organization for societies working in information processing, IFIP's aim is two-fold: to support information processing within its member countries and to encourage technology transfer to developing nations. As its mission statement clearly states,

> *IFIP's mission is to be the leading, truly international, apolitical organization which encourages and assists in the development, exploitation and application of information technology for the benefit of all people.*

IFIP is a non-profitmaking organization, run almost solely by 2500 volunteers. It operates through a number of technical committees, which organize events and publications. IFIP's events range from an international congress to local seminars, but the most important are:

- The IFIP World Computer Congress, held every second year;
- Open conferences;
- Working conferences.

The flagship event is the IFIP World Computer Congress, at which both invited and contributed papers are presented. Contributed papers are rigorously refereed and the rejection rate is high.

As with the Congress, participation in the open conferences is open to all and papers may be invited or submitted. Again, submitted papers are stringently refereed.

The working conferences are structured differently. They are usually run by a working group and attendance is small and by invitation only. Their purpose is to create an atmosphere conducive to innovation and development. Refereeing is less rigorous and papers are subjected to extensive group discussion.

Publications arising from IFIP events vary. The papers presented at the IFIP World Computer Congress and at open conferences are published as conference proceedings, while the results of the working conferences are often published as collections of selected and edited papers.

Any national society whose primary activity is in information may apply to become a full member of IFIP, although full membership is restricted to one society per country. Full members are entitled to vote at the annual General Assembly, National societies preferring a less committed involvement may apply for associate or corresponding membership. Associate members enjoy the same benefits as full members, but without voting rights. Corresponding members are not represented in IFIP bodies. Affiliated membership is open to non-national societies, and individual and honorary membership schemes are also offered.

HOME-ORIENTED INFORMATICS AND TELEMATICS

Proceedings of the IFIP WG 9.3 HOIT2005 Conference

Edited by

Andy Sloane
Professor of Telematics, University of Wolverhampton
United Kingdom

Springer

Library of Congress Cataloging-in-Publication Data

A C.I.P. Catalogue record for this book is available from the Library of Congress.

Home-Oriented Informatics and Telematics Edited by Andy Sloane

p.cm. (The International Federation for Information Processing)

ISBN: (HB) 0-387-25178-2 / (eB00K) 0-387-25179-0 Printed on acid-free paper.

Printed in the United States of America.

9 8 7 6 5 4 3 2 1 SPIN 11402985 (HC) / 11403036 (eBook)
springeronline.com

Preface

This volume is a collection of the papers presented at the HOIT2005 conference held in York, U.K. April 13-15, 2005. The conference was organized by IFIP Working Group 9.3 (Home-Oriented Informatics and Telematics).

The papers collected here are grouped according to the various strands of the conference namely:

Assistive technology, smart homes, home technology, memory aids, home activity, appliance design, design methodology, time, space and virtual presence, social and ethical aspects and home activities.

The placing of papers in the volume is random. The range of topics shows the growth in interest in this area of study and associated topics.

Andy Sloane,
Editor

Contents

Acknowledgments

The hard work of IFIP WG9.3 in gathering and reviewing the papers collected here is gratefully acknowledged.

FLEXIBLE USE OF TIME TO OVERCOME CONSTRAINTS

A time-geographical discussion about power and flexibility

Elin Wihlborg
National Institute for Working Life, Norrkoping, Sweden and Department of Political Science, Luleå University of Technology, Sweden

Abstract: Flexibility is often a keyword for people working in project-oriented organisations. It definitely influences their everyday life, since they are expected to be available and engaged anytime and anywhere and often with different types of obligations. However, there are constraints limiting people's possibilities in every specific time-space. The time-geographical perspective provides concepts for analysing constraints, by identifying: capacity, coupling, steering or authority constraints. The aim of this paper is to discuss the time-geographical concept of constraints and analyse how they are overcame by a flexible use of time. The paper is based on two field studies. The first was conducted in twelve households and included in-depth interviews at two occasions and in between the household members wrote time-diaries during a week. The second was more extensive. Two focus groups, each with 24 persons, who all had written time-diaries, were interviewed. The conclusion of these studies is that different forms of constraints define the outcome of flexibility and that the use of time can be an expression of power.

Key words: Time-space, flexibility, power, constraints.

1. INTRODUCTION

Flexibility is often a keyword for people working in project-oriented organisations. It definitely influences their everyday life, since they are expected to be available and engaged anytime and anywhere and often with different types of obligations. Without clear and defined working hours and a single working place, there are always possibilities for choice - where, when and how you should fulfil your job-related obligations. The home is a time-space pocket given new meanings when hosting work related activities.

However, there are constraints limiting people's opportunities in every specific time-space. What activities that can be conducted depends on how the local time-space pocket is organised, what resources are available and accessible to the individual and which constrains s/he experiences in the actual situation. Daily life is very much characterized of routines. Routines are activity patterns that we conduct almost without reflecting about why and how. However, if there are many opportunities for different ways of conducting daily life, people are not bound by routines. Instead the individual makes choices of how, when and where to perform different things during the day. Such choices are very much a result of the individuals' power to form their own life. S/he becomes more aware of their activity patterns, since s/he has to consider different possibilities and clearly motivate their choices. In such situations the individual pays more attention to constrains, since they experience them more clearly when they are no longer hidden by daily routines.

Time-geography is focusing the individual's activity patterns in the time-space. Constrains are divided into three categories related to the individual's own capacity and co-ordination with others as well as constraints set up by restrictions that have the legal power to steer the individual.

For time-geographical studies a specific method of time-diaries has been developed to make the daily activity patterns, as well as spatial and social contexts, visible. Thus the embedding of activities in time space can be analysed from the patterns of what people do rather than what they say they are doing and both unreflected and conscious constraints can be analysed. However, since time-geography do not attempt to explain why the outcome in time space is as it is, I will here try to develop the understanding of power by dividing and specifying the constraints and briefly apply them to flexibility and the use of ICT.

2. METHOD AND MATERIAL

This paper emanates from two field-studies. The first was more intensive, since I there followed twelve households with a more in-depth approach. The second was more extensive since one person from each of the study's twenty-four households participated and they were interviewed in small focus-groups. In both studies I have worked with a combination of time-diaries and in-depth / focus groups interviews.

The method has been developed from the time-geographical time-diary studies (see among others Ellegård 1993; 1994; 1999; Friberg 1990; 1998;

Ellegård & Nordell 1997; Ellegård & Wihlborg 2001). This method is based on an ambition to find out what people do instead of what they say they are doing. By letting them fill in simple diary sheets answering the questions: where, when, what they do, together with whom and some comments, the researcher gets a picture of their activity patterns, social interaction and movements. The method is based on time-geography, which combines the dimensions of time and space and sees them as inseparable. A crucial issue for the founder of the time-geography, Torsten Hägerstrand (se among other 1974; 1985) was the ambition to develop a notation system, which could grasp and combine observations in its time-space context (Hägerstrand 1974:88). Thereby time-space interaction was illustrated as trajectories where the map is the base (horizontally) and time develops upward (vertically). The time-diaries are seen as field notes and are converted into expressive illustration of the perspectives: time-space, time-activity and time-social contacts. The analytical tool for the time-activity dimension has been developed by Ellegård and Nordell (1997) and the different types of activities are categorized into seven main spheres. Each sphere includes about 100 activities combined in different groups. The main spheres of activities are: Care for one self, Care for others, Household care, Recreation and reflection, Movements and transportation, Procure and prepare food, Gainful employment and School work.

The complex patterns which are made visible through the time-space notation system also open up for comparison between households over time. However, the illustrations of what the informants have noticed in their time diaries say very little about the reasons behind the different activity patterns, social contacts and movements. Therefore we have found it essential to combine the time-diaries with different forms of interviews (Westermark 2003; Nordell 2002; Wihlborg et. al.2003). I made two interviews with each household in their home. The first time I introduced the study and the time-diary and it provided me also with an on-site observation, mapping the basic context of the household, which was useful for making a relevant analysis of their time-diaries. The follow-up interviews were more reflective, beginning with spontaneous responses to the time-geographical illustrations of their days and then continuing with discussions about the reasons for their choices in specific situations. Mannberg (2004) has raised the demand for a triangulation of methods by using the first interview as a life-story.

Both field studies are based on this three steps interaction with the households/individuals: interview, time-diary and reflective interview. The households in the in-depth field-study were chosen to illustrate different degrees of flexibility in their paid work. Since all these interviews have taken place in the home of the participants I did bring food (dinner or lunch depending on time of the day).[1] This group have totally noted about three thousands activities in their time-diaries.

The participants in the second field study had very diverse life situations, since they were students taking part in a part time course at the university. Some of them worked fulltime at the same time, others were full-time students whereas others had part-time employment or were self employed at the same time. However, they all had to be flexible to fit in the course which was given on different days and times and demanded a high degree of own work. These differences in life situation of the participants made the discussions in the focus-groups very dynamic. Each focus-group included eight persons and the groups were set up to be different regarding gender, age and household composition. During the first meeting we discussed the participant's expectations and their experience of being flexible enough to fit the university course into their daily life as well as some practical matters like how to deal with the time-diary. In their preparation for the second meeting they were asked to write down, based on their time-diaries, some reflections regarding: use of information technology, physical movements, and situations where they considered themselves as flexible. These reflective papers then formed the base for our second focus-group interview.

It is difficult to evaluate if the sample of households and individuals is scientifically valid since it is indeed a very qualitative study. The introduction of the study was based on quite vague research questions supported only by the potential benefits of the method and without any clear hypothesis. Therefore I could neither define factors that could be regarded as dependent and relevant nor could I select households that were representative without knowing what they should represent.

3. TIME-SPACE AND IT'S CONSTRAINTS

Time-geography is not a theory as such. It is rather an ontological perspective, suggesting that everything takes place and time, regardless if it is a person or a thing, and that humans and their contexts (physical, social and mental) are inseparable (Hägerstrand, 1974; 1985).

[1] This should also been seen as a small compensation for their participation since they did not get any economic compensation.

3.1 The time-geographic perspective

Time-geography has developed from geography and therefore the physical world and the human utilization of it, is in focus (Hägerstrand, 1993). Time-spatial relations between phenomena of different kinds are important in time-geography (Hägerstrand, 1985). There is a co-existence in time and space, since everything *takes place* somewhere on the earth's surface, and *takes time* to happen. These essential observations are the ground for the time-geographical basic *processes orientation* (Hägerstrand, 1974; Lenntorp, 1976). The process approach of time-geography underlines the importance of not only the events going on, but also of the existence of relics of past events and of seeds for future events immanent in present ones (Hägerstrand, 1974). The present *now* is continuously transformed into *past* situations. From an individual perspective this process includes the overcoming of constraints and an identification of what opportunities these offer depending on what has been done and what the plans are for the future.

3.2 The Problem of Power

Time-geography has been criticised for not making power relations visible. But as I stated in the discussion about methodology, time geography has so far mainly been concerned with issues like what is happening, when and where, rather than with the question of *why*. Therefor a time-geographical analysis usually stops with a description of the outcome in time-space without exploring the reasons that could be provided by an analysis of power. Giddens formulates this critic when he writes:

> ... time-geography involves only a weakly developed theory of power /... / both vaguely formulated and invoke a zero-sum conception of power as a source of limitations upon action. (Giddens 1984:117)

Hägerstrand's weak and actually quite uninteresting discussion of power can be seen as a consequence of the fact that he was satisfied when he had established the notation system. (The notation) This system was his main aim – not to explain the reasons behind what is represented in the notations of time-space. However, now that the notation system is there, we must consider *why* time-space is organised in the way it is. Thus I will try to develop Hägerstrand's concepts of constraints further by opening for a discussion about power. I will do this by dividing and defining the concepts of constraints that Giddens points out as being the ones closest to the discussion of power in time-geography, when he states that:

> If power is conceived as generative, on the other hand, the 'constraints' of which Hägerstrand speaks are all modalities for the engendering and sustaining of structure of domination. (Giddens 1984:117)

Hägerstrand himself makes a useful distinction between capacity constraints related to the individual. But, when it comes to the individual's more complex relations and thereby also to the compound constraints of relations to others in a social and physical context, he only provides a vague explanation. Therefore I will here try to show how the structures of dominance are articulated as consequences of power in time-space.

3.3 Constraints of our capacity

Hägerstrand (1985) begins with the individual person in her time-space context and focus on her capacity to act limited only by the most basic capacity constraints. Capacity constraints are, however, everything that limits the individual's actions, and that could be deduced to the capacity of herself or the tools she posses. There are three different types of capacity constraints. First the *biological capacity constraints*, such as the need for rest and food at certain intervals and sometimes even medical restrictions like having a cold for a short period of time or life time capability constraints by a disability. The biological limits are integrated with the body and are related to the nature of human beings. Also bound to the human being, but much more different between individuals, are the *mental or intellectual capacity constraints.* These are limiting her capacity to understand and cope with the information given, since knowledge (both theoretical and tacit) is restricted. Over a life time individuals continuously develop, more or less, mental and intellectual capacity in order to be able to overcome these constraints. We learn to speak, read and write and how to fit in to different social situations. Mental and intellectual capacity constraints are more contextually dependent[2] than are the biological capacity restrictions. The last form of capacity constraints, that Hägerstrand (1985) identifies, is the *spatial capacity constrains,* which limit our reach in time-space. For example, one can decide to sit on a chair next to the computer (not reaching any further in the real space (not virtual spaces)) or walk to the café (reach a spaces where other needs could be met) and if using any tools (bike, car, bus or...) less time is needed to reach the same positions. By having access (depends on coupling constraints) to tools for travelling or communication, like telephone

[2] We all have the experiences of not understanding a social context that we have experienced, for example being a tourist and not understanding the native language and cultural codes.

and e-mail, the spatial capacity is increased and the person can reach further in space in the same time.

3.4 Coupling constraints

Coupling restrictions is everything that limits the individual's co-ordination with other people and physical artefacts. Dealing with coupling constraints is a process of co-ordination. In the notation system the connection between people and artefacts is obvious and very often in focus. Thus it is important to separate coupling constraints into two groups covering coupling to *other humans* and to *physical artefacts* respectively, since other humans have the power to manipulate our coupling constraints, whether these can facilitate the coupling or not. The powers expressed by humans are deliberate.

Our possibilities to connect with other human beings depend on the other person's willingness to be connected with. This is not the case when it comes to "dead things" as physical artefacts. In the latter case it mainly depends on our own coupling abilities, such as being able to catch a train. The train as such has no resources to constrain our possibility to catch the train. However, other human actors can facilitate our possibilities to grasp physical things.

When it comes to co-ordinating with other people, they have the power to decide, if they wish to facilitate our connection effort or make it more difficult. We can influence upon other persons coupling constraints by using power. Being late for a meeting forces the other party to wait (or give up waiting after a certain period of time). Then it is not just a question of coupling possibility but also of which of the parties concerned that has the power in this specific situation. The power to connect is essential for time-geography since it is very much an issue of usage of the time space.

3.5 Authority constraints

Steering or authority constraints are defined by Hägerstrand (1985) as everything that has a legal power to steer the individual's actions and thereby limit her action space. Authority constraints are controlling different actors' access to time-space and thereby the power of these actors is obvious. The authority constraints are limiting our access to the space and thereby also our use of time, since they define which time-space that is accessible for us (Åqvist 1992). Opening hours for shops and other service places, for example, are authority constraints for our access to the activities that we can conduct there.

In order to understand the authority constraints we have to understand power. Since Hägerstrand's conceptualisation of power is vague, I will deal with the complex discussion of power by applying three different broad conceptualisations of power. This is a huge theoretical discussion and I will not be able to make a full coverage neither in this paper nor in a longer profound discussion. The following should thus be seen as a first attempt to deepen the time-geographical problematization of power.

The most basic way of understanding power was put forward by Dahl (1957) stating that A has power over B, when s/he makes B do what B otherwise would not have done. This is an explanation of power that does not consider the context and regards the parties as equally able to negotiate. It is only useful for explaining really simple situations and therefore it can be used to explain situations clarified revealed by the time-geographical notation system since it disregards the relation as such. Thus authority constraints can first of all be seen as those brought about by other *individuals' power* over us in a specific pocket of time-space.

There are, however, no such equal relations were power can be freely negotiated. The actors always have access to different resources, which they use in order to get what they want. The actors in a specific power situation also have a history of former experiences and desires as well as access to resources, which can give them the power in the specific situation. The use of resources have to be legitimate; otherwise there will be legal reprimands of using the resources. There are of course always risks that someone should use their resources in a non-legitimate way such as theft or violence, but to keep it simple. I'm disregarding such alternative possibilities in this paper. The legal use of resources to gain power provides opportunities for *legally* based *authority constraints* towards others. What is legal in a society, is determined by those of it's institutions that are structuring it's organisation (see among others Giddens 1984). Here power is the complex inter-relation between actors, their resources and their utilisation of institutions and social construction of resources.

Institutions can be both formal, such as legal rules, and informal, such as cultural norms and praxis (North 1990). Informal institutions are also considered as power in the society, but more implicitly. For the conception of constraints in this meaning we have to turn to Foucault's more discursive understanding of power. This is a form of power that regulates and controls human thoughts and behaviour and it is thereby the product of informal institutions in the society. Foucault argues that power just can be exercised and exists in action (Kelly 1994). In meaning no one can per se have power. Applying this perspective of power to the discussion of constraints, we have to be aware of constraints expressed through actions legitimated by *cultural norms* and *dominating discourses* seen as informal institutions. Power, in this meaning, regulates and controls human behaviour. The outcome of power is made visible at first through the human being's pattern of activity, which is in focus for the time-geographical notations system.

3.6 Summing up the constraints

The capacity constraints are found within the individual herself and can be expressed as the power her capacity has over her possibilities to act. Coupling constraints are either dependent on possibilities to connect to others or, if the other is a human being, that person's expression of power through her willingness and ability to connect. Finally, the authority constraints are defined by other individuals and social structures, legal and discursive.

Table 1. Summary of my view on different forms of time-geographical constraints

Capacity constraints	*Biological* capacity constraints *Mental and intellectual* capacity constraints *Spatial* capacity constraints
Coupling constraints	Constraints coupling to *other human beings* Constraints coupling to *physical artefacts*
Authority constraints	Authority constraints expressed by other *individuals power* over us Authority constraints expressed by *legal rules* (formal institutions) Authority constraints expressed in actions by *cultural norms and dominating discourses* (informal institutions)

The aim with this division into different forms of constraints is to show how power is an aspect of the constraints. These constraints are limiting access to resources and individuals are using time and other resources to prevail over them. I will now briefly apply this view on power and constraints to analyse my time-geographical field-works on flexibility.

4. TO BE FLEXIBLE AND/OR HAVE FLEXIBILITY – A QUESTION OF CONSTRAINTS

Flexibility can both be a personal characteristic and also a characteristic of the situation in which the individual acts. In other words you can *be* flexible and you can *have* flexibility (Bekkengen 1999). These two aspects of flexibility are in different ways limited by (the different) various constraints.

Flexibility is often seen as a positive quality or characteristic of a person, but being flexible is also demanding (Blocklehurst 2001). If flexible means 'not being bound to routines' it takes a lot of time and other resources to make new decisions continuously. If being flexible means 'to have competence to do different things' then there is also a demand for resources.

4.1 Capacities to be and have flexibility

To be flexible is a personal capacity in itself and the participants in my studies showed that they had different built-in personal capacities to work in a flexible manner. Many of the part-time students, who had to take on the course in order to keep their jobs, expressed biological capacity constraints. They did not manage to study in the evenings after a full day of teaching themselves. Their bodies had the power (biological and mental capacity constraints) to prevent them from working any more. They expressed this as an obvious but, for many of them, unexpected constraint.

To be flexible they also spent remarkable amounts of time commuting, especially the days of the course. Travelling is a concrete example of handling spatial capacity constraint and it takes time to overcome the spatial capacity constraints. Their possibility to have spatial flexibility was constrained by where they lived. The reallocation of time from other activities to travelling is an expression of power since they thereby also defined their priorities. ICT was sometimes used to overcome the spatial capacity constraints, but as shown in many other studies, this could not fully compensate face-to-face interaction – there are other intellectual capacity constraints of communication when mediated. Technology can make people more flexible, but communication is constrained when mediated. Therefore they have to put efforts into the development of resources such as routines for communication, trust and technical competence. It takes time to develop such skills. Competence for communication is culturally constructed and therefore highly contextual as well. Learning is a way to handle the capacity constraints of the technology and thereby to enlarge the power to manage technology.

4.2 Flexible coupling

The capacity to connect to other is essential and the participants who worked from home showed how their flexibility improved their opportunities and encouraged them to couple with other things and people. Working from home using ICT gave flexibility to reach domestic equipment during work-hours. This became obvious from the time-geographical notation based on their time-diaries, where short periods appeared in the column for "household work" breaking up the main activity: paid work. Typical household work project was to handle laundry including activities such as starting the washing machine, emptying, hanging and ironing. Thereby the person could couple to the apparatus at home, only restricted by a weak coupling constraint to reach an artefact as compared to if s/he had been working away from home. Making the laundry is an example of a project that takes more time than it needs attention from the person conducting it and therefore flexibility allowed it to be integrated with the paid work. One woman paid special attention to her management of these coupling constraints and saw the benefit of flexibility when comparing with the need to do all the laundry during weekends. She did not save time, but integrated activities and thereby she did not have to spend time neither to overcome spatial capacity constraints nor to overcome coupling constraints regarding artefacts.

Couple constraints related to other human beings and to organisations associated with the job can increase when tele-working. The informants in the intensive study had a need for technology, like using e-mail and telephones not just for specific information, but also for social reasons like saying good morning and telling colleagues that I'm working today, all as if they were doing some small talk in the office. Here coupling constraints could occur, however, when those being in the office didn't appreciate these contacts. Receiving a reply to a "good –morning" mail in late afternoon was almost insulting and gave a feeling of being excluded from the group spirit by the power expressed by the colleagues.

However, coupling to other people and to organisations other than those related to the work, were often facilitated by flexible working hours. Coupling to friends and even to services became much easier when they had the flexibility to spend the time as they wanted, as compared to working seven to four. Access to flexibility decreased the strength of the coupling constraints, since the informants could organize their use of time as they wanted.

4.3 The power to be flexible and to have flexibility

Everyone who expressed that they had experienced flexibility saw it as a power resource in the relations at the workplace. They could define their own action space and time-use. The ones in high positions within the company or the self-employed could define themselves what to do. They had power to administrate their own days and time – they were empowered. The flexible tele-workers had the power to do what they wanted in relation to others; they could prevail over authority constraints imposed upon others. They also had legal opportunity to work when and where they wanted and in addition cultural norms strengthened these opportunities, since there is a discourse between positive values and flexible work. This positive discourse gave them power when acting in relation to others since this is considered a desirable organisation of everyday life. Thus all aspects of the authority constraints related to their work were weak and they had all immense flexibility. On the other hand the legal authority constraints on the substance of work were, according to many of those with a flexible work, regarded as weak. This since they seldom had a clear agreement about the amount of work that was to be made and that they actually demanded more legal power from their employer in order to be able to feel satisfaction in their work. None really expressed any power to make them work more and spend less time on leisure activities. But through their use of time the discursive power is expressed.

On the other hand the shift-time workers lacked the power to overcome such constraints. They had no power to define their working hours themselves and the authority constraints were strong. Instead they needed a supportive home context to be flexible enough to come to work on a given schedule. Thus they had to have personal power to make others keep up at home for them when working odd hours, for example: taking care of children in the mornings of one week and at nights during the other. This is an authority constraint they practiced towards other individuals. They developed power to construct authority constraints on family and friends in order to make their everyday life flow smoothly. Their actions are outcomes of the strong and dominating discourse of paid work and the norm of having a job. Therefore they could keep up their power in relation to others, since the discursive power supported their organisation of everyday life.

5. IT TAKES TIME TO PREVAIL OVER CONSTRAINTS – CONCLUDING REMARKS

The participants made more or less conscious and more or less flexible use of time for handling daily constraints. Time is no constraint for household activities and choices. They all have 24 hours a day, and time is thus no variable constraint. It is rather the amount of activities people try to press in within their limited time-space that gives an impression of time as a constraint. Time can rather be seen as an integrated part of all the constraints, not at least since time is used to overcome the constraints. Time is used as a flexible resource that can open for many different ways of fulfilling a specific project by setting together different activities (Fitzpatrick

2004). Time itself is often enough to overcome the basic capacity constraints. You can sleep, study or walk far away to prevail these constraints. But time has to be combined with other resources, which are often extracted from artefacts or social institutions in order to overcome the more complex authority constraints. These authority constraints and the power they articulate also vary between different time-space pockets. The construction of power is always contextual.

The late modern society characterized by globalisation, consumption and 'the new economy' is aiming for the liberation of the individual person from everything that can limit her creativity and free choices of life (Jørgensen 2004). Therefore the discursive authority constrains are less obvious in the late modern era. This means that the possibility of having flexibility is greater. The more complex the society is the more tools are within reach for the individual. Thereby her capacities can develop in many different ways. However, since all human beings have these possibilities, the individually based authority constraints will increase whereas the discursive authority constraints, on the other hand, will allow more differences and thereby open up for flexibility without constraints.

Finally we must be aware of that there are always compromises made in view of the fact that we have to live with lots of constraints. There are compromises based on how flexible you can and will be and how much flexibility you have and want to use. Possibilities to *be* flexible are mainly restricted by capacity constraints. To *have* flexibility is rather an issue of authority constraints. The coupling constraints are limiting both the individual capacity and the importance of the authority constraints. The ambition to handle the constraints depends on the individual's personal sense of what is a good life.

REFERENCES

Bekkengen, Lisbeth (1999) "Män som "pappor" och kvinnor som "föräldrar"". In: *Kvinnovetenskaplig Tidskrift*, vol 20, nr. 1 pp. 33-48.

Blocklehurst, Michael (2001) "Power, Identity and New Technology Homeworks: Implications for 'New forms' of Organizing". In *Organization Studies*, may 2001.

Dahl, Robert A. (1957) "The Concept of Power" *Behavioral Science/ Journal of the Society for General Systems Research*. No 3: 201-215.

Ellegård, Kajsa (1993) *Olikadant. Aspekter på tidsanvändningens mångfald*. Occasional Papers 1993:4. Kulturgeografiska institutionen. Göteborg: Göteborgs universitet.

Ellegård, Kajsa (1994) *Att fånga det förgängliga. Utveckling av en metod för studier av vardagslivets skeenden*. Occasional Papers 1994:1. Kulturgeografiska institutionen. Göteborg: Göteborgs universitet.

Ellegård, Kajsa (1999) A time-geographic approach to the study of everyday life of individuals – a challenge of complexity. In: *GeoJournal* vol 48 no 3 1999, Special issue on

"Actors, activities and the geographical scene, Studies on time-geography, mobility and gender."
Ellegård, Kajsa och Nordell, Kersti (1997) *Att byta vanmakt mot egenmakt*. Stockholm: Anderssson & Skyttemo.
Ellegård, Kajsa och Wihlborg, Elin (2001) *Fånga vardagen. Ett tvärvetenskapligt perspektiv*. Lund: Studentlitteratur.
Fitzpatrick, Tony (2004) "Social Policy and Time" In: *Time & Society*, vol. 13, pp 197-219.
Friberg, Tora (1990) *Kvinnors vardag. Om kvinnors arbete och liv. Anpassningsstrategier i tid och rum*. Lund: Lund University Press.
Friberg, Tora (1998) *Förflyttningar, en sammanhållande länk i vardagens organisation*. (KFB-rapport 1998:23). Stockholm: KFB.
Giddens, Anthony (1984) *The Constitution of Society*. Cambridge: Polity Press.
Hägerstrand, Torsten (1974) "Tidsgeografisk beskrivning – syfte och postulat". In: *Svensk geografisk årsbok*, 86-94.
Hägerstrand, Torsten (1982) "Diorama, path and project". In: *TESG Tijdschrift voor economische en sociale geographie. Journal of economic and social geography*. 1982 lxxiii nr 6.
Hägerstrand, Torsten (1985) *Time-Geography: Focus on the Corporeality of Man, Society, and Environment*. New York: The United Nations University.
Jørgensen, Carsten René (2004) *Psykologin i senmoderniteten*. Stockholm: Liber AB.
Kelly, Michael (ed.) (1994) *Critique and Power*. Cambridge, MA: MIT Press. Including Foucault, Michel "Two Lectures".
Lenntorp, Bo (1976) *Paths in space-time environments. A time-geographic study of movement possibilities of individuals*. Lund: Gleerup.
Mannberg, Mariann (2004) "Tidsgeografisk triangulering. Om tidsgeografin som kombinerad samhällsvetenskaplig metod". In: Ellegård, Kajsa & Wihlborg, Elin *Tidsgeografiska perspektiv*. Tema T working paper. Linköping: Linköpings universitet.
Nordell, Kersti (2002) *Kvinnors hälsa. En fråga om medvetenhet, möjligheter och makt*. Göteborgs universitet.
North, Douglass C. (1990) *Institutions, institutional change, and economic performance*. Cambridge: Cambridge University Press.
Tietze, s. (2002) "When "works" come "Home": Coping Strategies of Teleworkers and their Families." In: *Journal of Business Ethics*. 41, pp 385-396.
Westermark, Åsa (2003) *Informal livelihoods: Women making and reflecting upon their biographies. A time geographic analysis in urban Colombia*. Dept of Human and Economic Geography. Göteborg: Göteborg university.
Wihlborg, Elin, Jessica Rahm och Kajsa Ellegård (2003) *Konvergerande teknik kommer hem*. (Tema-T Arbetsnotat 272). Linköping: Tema Teknik och social förändring, Linköpings universitet.
Åquist, Ann-Cathrine (1992) *Tidsgeografi i samspel med samhällsteori*. Lund: Lund University Press.

COOK'S COLLAGE

Déjà vu Display for a Home Kitchen

Quan T. Tran, Gina Calcaterra, Elizabeth D. Mynatt

College of Computing, Georgia Institute of Technology, Atlanta, GA USA

Abstract: Many people regularly multitask while cooking at home. Juggling household chores, reusing limited kitchen utensils, and coordinating overlapping cooking times for multiple recipes can cause frequent task switching and simultaneous task monitoring while cooking. As a result, the cook occasionally loses track of his cooking progress especially when determining which ingredients have already been added, counting multiple scoops of an ingredient, and keeping watch of cooking times. People compensate for these memory slips by devising memory strategies or deferring to memory aids with varying degrees of success. In this paper, we present a novel memory aid for cooks called Cook's Collage. We describe how the system constructs a visual summary of ongoing cooking activity. Then, we report a task simulation study evaluating the effectiveness of Cook's Collage as a memory aid. We argue that a memory aid is helpful only if it is balanced correctly with a complementary memory strategy and only if the accuracy of the memory aid is trusted. Lastly, we discuss how the six design features of the Cook's Collage suggest a general framework for memory aids in the home, which we term déjà vu displays.

Key words: Memory aid, appliance design.

1. INTRODUCTION

Memory slips can occur while cooking at home for many reasons. First, the lack of visual cues may make determining cooking progress problematic. Consider mixing different ingredients of similar color and texture such as flour, baking soda, baking powder, and sugar. Remembering which ingredients have been added and how much of each has been added can be problematic. All the mixed ingredients are available and visible, but they are not easily distinguishable. Second, failures of remembering self-performed actions can result from routinized tasks that occur automatically without conscious awareness, from memory decline with increasing age, or from confusion between already performing an action and intending to perform an action [4]. Lastly, interruptions and multitasking can create memory slips by overloading memory demands. Juggling household chores, reusing limited kitchen utensils, and coordinating overlapping cooking times for multiple recipes can cause frequent task switching and simultaneous task monitoring while cooking. Gillie and Broadbent [6] showed that interruptions are more disruptive the more they are similar to the interrupted task, the more complex they are, and the more time they consume.

The cook occasionally may forget and question his cooking progress. "Which ingredients have already been added?" "How many scoops of an ingredient have been added?" "How long has this item been cooking?" Not knowing the answers can ruin recipes. Most people compensate for their memory slips with memory strategies and memory aids with varying levels of success. Identifying those who could benefit from memory aids may be paradoxical since [18] shows that those who commit memory slips are also often oblivious of their memory slips.

We present Cook's Collage, a novel memory aid for general cooking that constructs a visual summary of ongoing kitchen activity. We have demonstrated this memory aid to numerous visitors including researchers, local families, college students, elderly adults, and members of the press and media throughout the past three years. Cook's Collage receives enthusiastic support from many people who feel that they have memory slips while cooking and understand that these memory slips consequently cause problems. In this paper, we detail the design features of Cook's Collage. Then, we report a task simulation study evaluating the effectiveness of the memory aid. Lastly, we discuss how the six design features of the Cook's Collage suggest a general framework for memory aids in the home, which we term déjà vu displays.

2. COOK'S COLLAGE

Cook's Collage is a novel memory aid that constructs a visual summary of ongoing kitchen activity. The system is based on six design features. In this section, we detail the system specifications of the Cook's Collage with respect to the six design features. Afterwards, we revisit the design features, discussing their rationale and implications for home memory aids in general.

2.1 Distributed Cognition, Capture, and Access

The system components are distributed and embedded within the home kitchen. Capturing devices (e.g., webcams) mounted underneath the kitchen cabinets as shown in figure 1b continually record video of countertop activity. This location hides the devices from casual view, rendering them inconspicuous and innocuous as shown in figure 1a. The system uses two cameras, the minimum number to provide overlapping spatial coverage of one corner of the kitchen countertops. Cook's Collage is presented on a LCD flat panel that hangs from an overhead cabinet, at eye-level with most cooks when standing. To be readily accessible, the display is positioned at the center of the kitchen triangle, a conceptual area understood by kitchen designers delimiting the three main components of a kitchen (i.e., refrigerator, stove oven, and sink). The LCD flat panel provides a slim encasing snugly infused within the cabinet fixture, physically presenting the output display as framed art or some other home artifact. With the physically dispersed components, the cook is able to cognitively distribute his memory load for the capturing cameras to be later accessed via the display.

2.2 Evanescence of Recent Activity

Cook's Collage aids short-term and working memory, presenting only the most recent activity and not an entire running history. The visual summary retains an evanescence of recent activity by continually updating the display and transitioning the shown actions. As a new action is performed, the system adds its corresponding information to the bottom right corner of the display, marks it as most recent by highlighting a yellow border, and shifts the previous actions one position left. The oldest action rests at the top left corner of the display. Once six actions have been performed, the next action causes the first action to transition off the display. Repeated steps like adding multiple scoops of an ingredient are considered sub-actions, so they transition differently. If the most recent action displayed is repeated, the new action replaces the previous action on Cook's Collage and its new numeric count is annotated appropriately. If a step repeats successively, its prior occurrences are compressed into gray bars (i.e., visual ellipses denoting repeating information without consuming space), and the new action with the accumulating count annotation replaces the previous one.

2.3 Juxtaposition of Displayed Information

The Cook's Collage displays a visual summary of ongoing kitchen activity by juxtaposing the six most recent actions in a relative temporal sequence.

The decorative film reel motif suggests a time flow, and the horizontal film reel borders cluster the rows of images together, suggesting a horizontal reading of the action sequence. Additionally, the 2x3 grid layout commonly used for comic strips suggest a left to right, top to bottom reading. For example, figure 2 retells that the user added the 29th scoop of sugar the last time he was here at the kitchen corner countertop. Before that series of sugar scoops, he had added two amounts of soda pop one after the other; before that, two amounts of ice one after the other; before that a series of sugar count ending with the 25th scoop; before that, a series of powder count ending with the fifth scoop; and before that, two amounts of water one after the other. The layout of the display is essential in interpreting the images.

2.4 Already Known Information

Retrospective memory entails remembering actions one has already performed. To support retrospective memory recall, Cook's Collage shows only already seen information. The memory aid uses raw video as visual cues, literally already seen by the user. However, repeated actions are visually ambiguous on video. The memory aid avoids potential ambiguity of visualizing repeated actions by collapsing the repeating action into a single visual representation with numeric annotation. This solution explicitly shows information that otherwise was implicitly already seen.

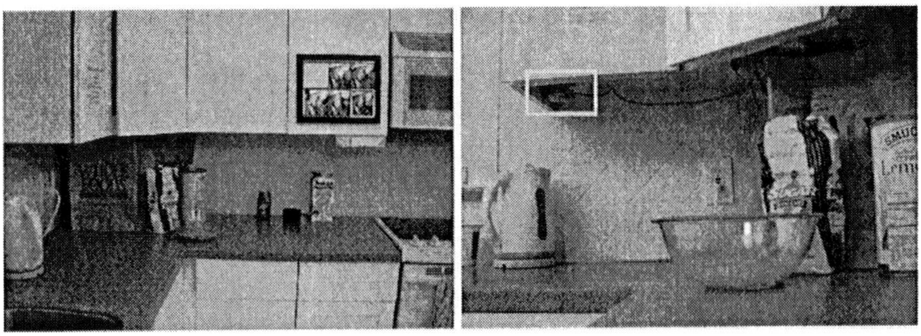

Figure 1. Kitchen setup at (a) eye-level view (b) close-up underneath view of cameras

2.5 Video

The video used for *Cook's Collage* consists entirely of close-up hand shots over the kitchen countertops. This allows the capture of necessary detail and avoids unnecessary occlusions. The positioned camera angles avoid an over-the-shoulder, surveillance view that could induce big brother privacy concerns and wrongly encourage "cooking show" theatrics uncharacteristic of everyday cooking. The use of unedited raw video accentuates the reality of the cooking experience and personalizes the cooking narrative.

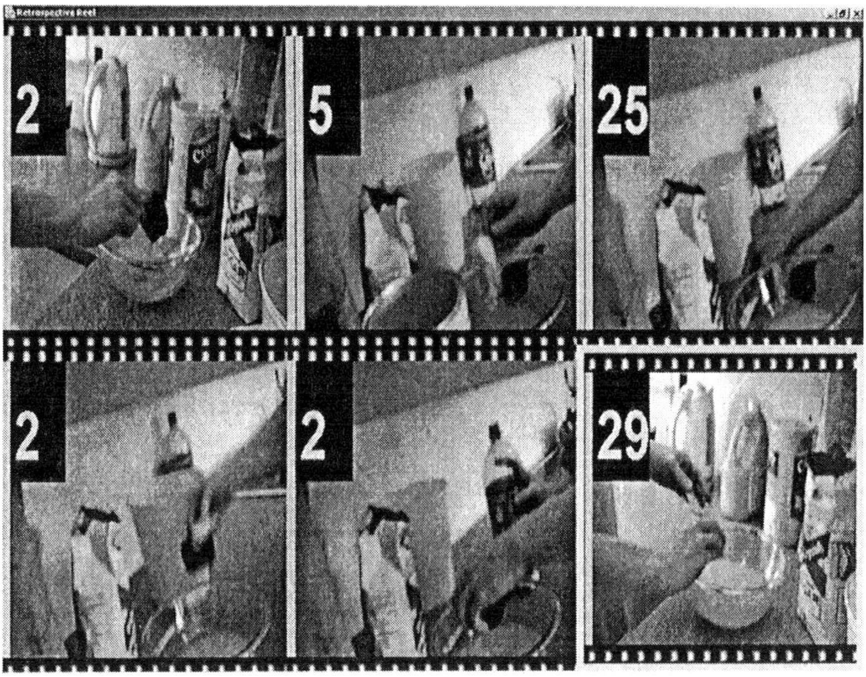

Figure 2. Cook's Collage display

2.6 Units

Cook's Collage represents a single step in a recipe as a unit of action, and shows each unit of action as a snapshot image in the display. Each image is displayed at 322x288 resolution, the minimum required for non-blurry images.

3. WIZARD OF OZ SYSTEM IMPLEMENTATION

The collection of images displayed on Cook's Collage is constructed in real time via a Wizard of Oz approach [9]. In a remote room, a human operator monitors a streaming assembly of image stills captured at a rate of one frame per second from which she hand picks a still image representing the user's action. Upon each image selection, the streaming assembly of captured images pauses so that the wizard can either associate the chosen image with its corresponding action and document her image selection heuristic or cancel her selection choice. Upon completion, the stream of captured images resumes.

Unfortunately, the image selection causes the system to pause, introducing variable delays into the system's update latency. Network routing between the distributed system components also imposes time latency between when an action is performed to when it is updated on the display. The wizard can commit human errors in interpreting the cook's actions from the streaming video and in operating the interface that builds the collage. Still, the time latency and human error of the current system may be no faster and no more accurate than the computational latency and low confidences of an automated perception system in the future.

In a previous version, Cook's Collage utilized various ingredient containers instrumented with RFID tags that sensed when ingredient bottles were being poured and automatically generated the collage summary. However, the sensing technology restricted the cooking behaviors we are exploring. Cook's Collage could leverage vision tracking, but we use the video medium as raw input. The captured video is not processed or synthesized in any way.

We select the wizard of oz simulation to permit evaluation of how the display designs affect possible user interactions without committing to a fully automated system. We believe Cook's Collage can be realized either by technical breakthroughs which will be years in the future, or by achieving the appropriate balance of human computer interaction, leveraging both human and computer abilities, which may be understood from our simulation studies. Currently, computational perception systems are unable to recognize the variability of everyday cooking that Cook's Collage supports. In a highly variable and dynamic activity such as everyday home cooking, sensing technologies lack precision to physically perceive individual tasks, and inference models lack accuracy to intelligently interpret overall activity. By studying how users interact with the Cook's Collage, our findings suggest that a fully automated system may not be necessary because people willingly adjust their memory strategies and accommodate the memory aid in a collaborate effort to improve overall memory recall.

4. RELATED WORK

Albeit office groupware, [12] is used in a similar manner as the cook's collage. As a team member arrives late to a running conference meeting, he scans video segments from the ongoing meeting to be quickly briefed of the arguments leading up to the current discussion. While there are some memory support systems for the home (e.g., [5,11]), they investigate software infrastructure to support large-scale archives of multimedia. We have yet to find another memory support system for cooking. Various kitchen gadgets such as egg timers or cooking appliances that turn off after a preset time assist with one of the three potentially problematic cooking tasks we identify, keeping track of cooking times. We also addressed an alternative approach to keeping track of cooking times with an earlier design of Cook's Collage [16]. The other cooking tasks, remembering which ingredients have been added and counting multiple scoops of an ingredient, are not currently supported by memory aids but rather by memory strategies that are prone to error. We note examples of these memory strategies in our experimental findings.

There are many systems that help novice cooks learn new recipes. Although also kitchen applications, the focus of these systems is primarily to assist the cook with navigation through recipes rather than Cook's Collage's focus of supporting memory recall. For example, eyeCook [13] tracks the cook's eye gaze to infer which recipe instruction or ingredient description to read out loud next. Another kitchen system [14] uses foot panels along the bottom kitchen cabinets for the user to explicitly navigate through recipe instructions.

5. EXPERIMENTAL EVALUATION METHOD

A task simulation study is defined as "a laboratory memory test that bears some resemblance, at least in stimulus content, to an everyday memory experience" and is valued as the "best situation" in terms of research with high ecological validity (from using cues or stimuli that have potential utility in real-world environments) and high generalizability (from the laboratory control and potential for programmatic replicability) [17]. A task simulation study was used to evaluate the effectiveness of Cook's Collage as a memory aid. We designed a dual task scenario that consistently produces memory slips for cooks. This experimental setup enables comparison of the cook's memory performance with and without Cook's Collage.

5.1 Dual Task

We designed a dual task to overload working memory. We drew inspiration from our previous experiment's results that rank having an ongoing conversation while performing a separate task as most difficult and annoying to manage [15]. We chose a language learning task to provide a sufficient distraction to the cook. Though things such as spontaneous interruptions may be more common distractions in the kitchen, they are more difficult to control. We thus decided to use a multitasking situation that is similar to activities cooks perform such as engaging in conversation in the kitchen, talking on the phone, and interacting with their children. Though language learning itself is not a typical activity cooks engage in while multitasking in the kitchen, it does effectively induce cognitive load in the same way as the aforementioned activities. In addition, using the language task allows us to quantitatively measure performance. Directly measuring performance on the language task gives us a fuller understanding of how effectively participants use the memory support system. We counterbalanced for learning effect, language difficulty, and availability of the support system, to compare memory performance quantitatively and user experience qualitatively. All participants received four conditions in the following order: single task, dual task, single task, dual task. 11 participants first did the dual task with Cook's Collage as memory support, then did the same task without memory support. The other 11 participants did the dual tasks in the reverse order.

To encourage a dual task situation, we asked participants to try to pace their cooking task so that they would finish both tasks at approximately the same time. To avoid having the participants ignore the language task, we emphasized the language learning task as the primary task to be evaluated and the cooking task as secondary. No internal or external memory strategies were suggested in order not to prime the participants' behavior; however, they freely initiated and utilized whatever learning and counting strategies felt comfortable.

5.1.1 Primary Task: Language Learning

This task involved learning how to count to ten, five digits at a time in a foreign language (e.g., Vietnamese 1-5, 6-10; Chinese 1-5, 6-10). The learning task was accomplished by listening to an audio tape consisting of an opening welcome, statement overview, one complete counting sequence, instructions to start drilling by repeating each number after hearing a pairing of the English meaning and number pronunciation, four rounds of five number pairings, and finally a closing encouragement. The lesson spanned two minutes, looping once to total four minutes. The tapes contained a fairly consistent audio silence between digits allowing participants time to decipher pronunciation of foreign number and to repeat the digit aloud. The participants were not permitted to pause, rewind, or replay the preset lesson.

5.1.2 Secondary Task: Punch Preparation

The cooking task was to prepare punch. This task comprised of adding ingredients (e.g., 2/3 cup orange juice, 2/3 cup water, 5 scoops lemonade powder, 29 scoops sugar, 2/3 cup ice, 2/3 cup citrus soda) into a large bowl from a recipe posted on an overhead cabinet (Figure 1). Provided measuring utensils included an arbitrary sized scoop for the powder, a 1/3 measuring teaspoon for sugar, and 1/3 measuring cup for the remaining ingredients. Participants could add the ingredients in whatever sequence felt comfortable, and could freely rearrange countertop items excluding the punch bowl that was optimally positioned for the Cook's Collage cameras.

5.2 Participants

22 graduate students (14 male, 8 female) of ages 23-51 participated in this study. None of the participants learned how to count to ten in Chinese or Vietnamese prior to the study. 19 reported having previously used audiotapes to learn a foreign language (e.g., English, French, German, Italian, Spanish, Greek, Japanese, Russian), of which four reported multi-tasking more than 75% of the time while listening to language tapes. The majority of the participants reported multitasking more than 50% of the time while preparing food and drinks within their kitchen.

5.3 Cook's Collage Introduction

Prior to the experiment, each participant was given an overview of the Cook's Collage and the layout of the kitchen. The display features were debriefed, and a brief demo highlighted how updates of new images and number annotations occurred. The system limitations were also noted. Particularly, participants were told to expect variable time latency per display update. It is important to note that no hints about developing particular strategies for system use were given to the participants. Rather, the intent was to inform participants of the system features and limitations, allowing each participant to decide if and how Cook's Collage could be a useful memory aid. This debriefing constitutes the only introduction to the system that each participant received, thereby rendering each participant as a true first-time user of the system.

5.4 Units of Measurement

We quantitatively and qualitatively compare participant performances across the four conditions. For the language task, participant rated the pace of the tape drills and indicated which numbers they believed they learned and remembered, writing in reasons for numbers they believed they had not learned and remembered. A short oral quiz (e.g., reciting numbers in correct sequence and translating two random numbers) was administered to assess actual learning performance. For the punch task, we conducted semi-structured interviews to gauge user experience qualitatively. We used video record of the dual task to compare actual performance to participants' perceived performance. We used another video record to tally the number of times a user glanced at the display and the time duration of each glance.

6. EXPERIMENTAL FINDINGS

The experimental results produce further evidence that multitasking while cooking does cause memory slips, that memory slips are problematic for certain cooking tasks, and that many existing memory strategies people make use of can be faulty. We report how the cooks used Cook's Collage to cope with the taxing memory demands of certain cooking situations. We argue that a memory aid is helpful only if it is balanced correctly with a complementary memory strategy and only if the accuracy of the memory aid is trusted.

6.1 Did the dual task overload working memory?

Yes. Overall, participants recalled significantly fewer words in the dual task conditions than in the single task conditions, $t = 4.294$, $p < .001$. None of the 22 subjects rated the pace of the single task as too fast. 10 (45.5%) of the participants rated the pace of the task as "just right" on all four conditions. However, 11 (50%) of the participants rated a relative increase in the pace of the audiotape during the dual task conditions. One participant rated the pace of the audiotape on all four conditions as "too slow." Thus, any relative difference in the pace of the tape in the dual task conditions was rated as an increase most likely due to the increased cognitive load of the participants. No significant difference was found between the number of words a participant recalled on the first single task and the second single task, $t(21) = -1.164$, $p = .257$. We can conclude that Chinese and Vietnamese are of relatively equal difficulty level for individuals who do not speak an Asian language.

6.2 Did participants adopt successful memory strategies to aid them in the dual task?

Participants employed a variety of memory strategies in an attempt to externalize the information about the ingredients added and the number of ingredient counts completed, that they would otherwise hold in working memory. For example, some cooks used relative grouping of ingredients on the kitchen countertop to remind themselves which ingredients have been added and which have yet to be added. This however did not externalize information such as the number of counts of a particular ingredient that were added. In fact, most of the strategies participants adopted were not completely successful. As an example, some participants counted on their fingers to aid them in keeping tracking of ingredient counts. However, their timing was often problematic because the repeated numbers on the tape were easily confused with the ingredient count numbers, and this sometimes lead to counting errors. Other cooks used the drilling pattern in which numbers are presented on the language learning task to pace their ingredient additions. This strategy worked very well for this dual task, since the grouping of the numbers on the tape was in fives as well as the grouping of most of the ingredients. However ingenious though, this strategy cannot be generalized to other multitasking scenarios, and thus cannot be considered an effective way for the cook to externalize working memory information into the world in a typical dual task situation. Thus, though people did adopt a variety of memory strategies to aid them in multitasking while cooking, these strategies were often faulty and would not generalize well to all multitasking situations that a person may find himself in while cooking.

6.3 Did first-time users find the memory system useful?

Yes, the memory system can be useful if the user adopts a successful strategy for adding the ingredients and using the memory aid during the dual task. Here we define a strategy as a way of utilizing the memory aid to help in the punch preparation task, allowing the participant to concentrate more on the primary task. We found that participants who had no practice performing a dual task before using the memory aid (i.e., with memory aid, without memory aid order of conditions) had a harder time developing a successful strategy for using the system, and in some cases chose not to use the system at all but rather relied on their own personal memory strategies. 63.6% (7/11) of the participants who received the dual task condition without memory support before receiving the memory support condition did use the memory aid and even developed a successful strategy of interleaving ingredients when they had the memory aid available to them. This strategy was effective because it allowed the cook to add an arbitrary number of scoops of an ingredient, and then add more scoops later on based on the output displayed on the collage for that ingredient. They let the system first "catch up" with them and then determined what steps remained, rather than keeping track of the counts themselves and adding all of an ingredient at once. On the other hand, only 27.3% (3/11) of the participants who received the dual task conditions in the reverse order developed a strategy for using the system. Thus, participants who were already familiar with the dual task were better able to develop a strategy the second time they were put in a dual task situation. Many of the participants who had prior exposure to the dual task situation openly admitted in the post-task interviews that their own personal memory strategies used when they did not have the aid had been faulty, and they felt that using the aid would be a better alternative to relying on their own memory. This pattern was not observed with participants who had not had prior exposure to the dual task before having the option of using the aid. These participants tended to rely more on their own memory strategies, such as those discussed in a previous section, and did not see a need for the memory aid. In summary, practice with the dual task before being exposed to the memory support system made many participants offload their memory of the running list of used ingredients and the running count of multiple ingredients onto the aid rather than relying on their own memory strategies which had proved faulty in the previous dual task situation. They were also more able to develop a strategy for using the system to help them perform the dual task.

6.4 Following practice performing a dual task, can the memory system improve performance on the primary task?

Many of the participants who did offload their working memory of the running list of used ingredients and the running count of multiple ingredients actually improved their performance on the primary (language learning) task. The number of words recalled on the language quiz given after each condition was used as a quantitative measure of performance. 45.5% (5/11) of participants who received practice performing the dual task prior to introduction of the memory aid, either improved or showed consistent performance on the memory recall quiz. 4 of these participants actually changed their strategy from chunking ingredients (adding all scoops of a particular ingredient at once) on the first dual task condition to a strategy of interleaving ingredients on the dual task with memory support. We found that the participants who showed improved or consistent performance with the memory aid developed this strategy of interleaving ingredients to use the time latency of the memory support to their advantage. For example, one participant adopted a strategy that took advantage of the memory aid by 1) initially adding an arbitrary number of the 29 scoops of sugar, 2) moving on to those ingredients with a smaller number of steps, 3) referring to the display to see where she left off on the sugar count, and 4) adding the final scoops of sugar. Two participants actually improved their performance by adopting this strategy (e.g., P1, P26). On the contrary, we found that participants who maintained a strategy of chunking ingredients in both dual task conditions (e.g., P9, P22) showed no improvements in memory recall. Due to the latency and transitions of the display updates, chunking did not prove to be a very successful strategy for performing the punch task. Examples of both successful and unsuccessful strategies employed by 5 of the participants are shown in Table 1.

Table 1. Examples of memory strategies in the two dual task conditions.

Strategies adopted by participants in dual task conditions	***Without aid***	***With aid***
P1 – improved performance with memory aid	chunking (tap finger on counter after every 3rd scoop of sugar)	interleaving (add OJ, water, powder, arbitrary amount of sugar, ice soda, then refer back to display and add remaining sugar)
P26 – improved performance with memory aid	chunking (lost count at scoop 6, attempted to restart count, had to guess)	interleaving (add arbitrary amount of sugar, add water, OJ, look at display and add another arbitrary amount of sugar, add ice, soda, powder, look at display and add the rest of the sugar)
P11 – perfect performance on both conditions	chunking (counted sugar in increments of 10)	interleaving (add powder, OJ, water, ice, arbitrary amount of sugar, add soda, refer to display and add remaining amount of sugar)
P9 – decrease in performance with memory aid	chunking (counted on fingers, "stored the most recent count in my head")	chunking (add OJ, water, powder, add sugar, waiting for display to catch up, add more sugar, add ice, add soda)
P22 – decrease in performance with memory aid	Chunking (tried to make a mental note after each set of 5 scoops were added)	chunking (add all ingredients except sugar, add sugar last, overshot by 12 scoops)

Another interesting thing to note is that none of the 5 participants who improved or showed consistent performance rated an increase in the pace of the audiotapes on the memory aid task. 3 of the participants rated both dual task conditions as “just right” whereas the other 2 participants actually showed an increase comfort level with the pace of the tapes that they had rated as “too fast” on the dual task where they had no memory aid. This suggests that those who felt comfortable with the pace of the task showed improvements or consistency in memory recall.

6.5 Trust

The memory aid successfully allowed the cooks to offload the running list of used ingredients and the running count of multiple ingredients that are in working memory out into the world. However, Cook’s Collage was only successful when the cook actually trusted the accuracy of the system enough to use it. Results from this study indicate that many people felt they could trust the Cook’s Collage more than they could trust their memory. After performing the task, users were asked whether they trusted the system. Three of the five users who improved their performance when having the memory aid present the second time they performed the dual task reported that they trusted the aid more than their own memory. When asked if she trusted the memory aid, the first of these users commented, “Yes, definitely, I wasn’t as dependent on my own memory.” Another user commented, “I definitely trust it. It’s probably more accurate than my memory.” Yet another participant commented, “I made the decision early on to trust it, because it’s probably better than my own memory.”

7. DISCUSSION: FROM COOK’S COLLAGE TO DÉJÀ VU DISPLAYS

Based on the encouraging feedback from people who used Cook’s Collage, we explore other home activities that could also benefit from a memory aid. Possible household activities include taking medication, getting ready in the morning, and fixing household appliances. We revisit the six design features of Cook’s Collage to define a framework that we term “déjà vu” for memory displays. We conjecture how they contribute positively to the success of Cook’s Collage and discuss how to apply them in the design space for memory aids in the home.

7.1 Distributed Cognition, Capture, and Access

A powerful approach to addressing memory slips is to relieve working memory of keeping "knowledge in the head" by distributing respective information as "knowledge in the world" [10]. By making retrospective memory external, recalling self-performed actions transforms to recognizing information.

Cook's Collage directly uses this approach by capturing information about ongoing events and displaying this information visually within the environment to be accessed later by the cook as needed. Cooks can either offload information in working memory including the running list of used ingredients and the running count of multiple ingredients to the memory aid, or confirm the running counts of ingredients in their head with the running counts displayed by the collage. In similar means, people may complement or confirm their working memory for other household activities or in general with a déjà vu display.

7.2 Evanescence of Recent Activity

Short-term memory and working memory are used for activities that are short-lived and evanescent, yielding minimal long-term memory retention. *Déjà vu displays* summarize recent activity based on actions and not based on time. The action-based definition of recent activity complements the otherwise time-based description of evanescent activities common within the home.

Cook's Collage shows that an action-based history readily supports interruption recovery and task switching. The display answers the query "what did you last do here?", whether recent activity was within the previous minute if heavily active or within the prior week if activity was sparse. We believe that activities at home are continuous, without a definite beginning and end. Therefore, managing interruptions and task switching will be important for memory aids to support.

7.3 Juxtaposition of Displayed Information

Presentation of information is important. Juxtaposition of displayed information helps to compose or to isolate the units of information so that correct interpretation of the recorded narrative is possible. The composition of the display can either maintain structure or can be haphazardly scattered as a collage. Each déjà vu display may explore this range to leverage various memory strategies (e.g., temporal sequencing, spatial memory, semantic association).

Using a temporal sequence was fairly successful as a straightforward summary of past actions. This reinforces survey findings that mental retracing is the most frequently used internal memory aid by far for students (97% ever used, 23% use >1 per week) because of its advantages as a pure retrieval strategy, requiring no special encoding effort to use it [7]. The annotations for repeated steps and visual ellipses confused some cooks, who misinterpreted the overall display arrangement as a spatial layout. Cook's Collage could be alternatively redesigned with a different layout (e.g., [1]) under the déjà vu framework.

7.4 Already Known Information

Déjà vu is a memory recall phenomenon defined as "an impression of having seen or experienced something before; the experience of thinking that a new situation had occurred before." This experience epitomizes the memory slips we seek to address, because memory recall worsens by adding confusion between performing an action and intending to perform an action. The secondary definition of déjà vu is "a dull familiarity and monotony." The sense of familiarity (i.e., rich context inherent in already seen and felt personal experiences) from déjà vu inspires our scaffolding of retrospective memory support.

Our experimental findings show that cooking involves instances where cooks experience the former definition of déjà vu, especially when questioning which ingredients have already been added and how many scoops of an ingredient have already been added. Our exit interviews indicate that cooks experience the latter definition of déjà vu when reviewing Cook's Collage of their personalized cooking experience. In fact, study participants can quickly query Cook's Collage and easily interpret the informative details, whereas objective viewers study the memory aid at length and are frustrated with the information clutter. This reaction is similar to how daily commuters skim subway maps compared to how foreign visitors study them. There are many examples of both definitions of déjà vu experiences in everyday life that this framework can address.

7.5 Video

The video medium, captured live and unedited, maintains much of the original context, (e.g., first-person perspective, same visual modality) minimizing cognitive effort in interpreting the display of already seen information. The video medium is also rich in context and content. Moreover, the video format invokes an emotional and contextual identification akin to a déjà vu response to the captured personal experience much more than other formats (e.g., text, abstract graphs, charts) of event logging would.

Surprisingly, people reacted positively to the video used in Cook's Collage. The video was able to capture the necessary details of cooking to be effective as a memory aid without alarming cooks of privacy concerns of having cameras in the house. Our success using video reinforces similar findings [2,3] in the work place that video diaries increased the number of activities that were recalled relative to using only written diaries. We promote the use of video across déjà vu displays.

7.6 Units

To quantify and segment recent activity, the displayed information is parceled into incremental units. Within video medium, these units range from still image snapshots, to animated sequences of compressed video, to the original video replay.

Cook's Collage defines image stills as a video unit representing one cooking action. The memory aid presents a visual summary of image stills rather than presenting an interface to navigate through video replays of cooking actions. Our experiences using the Cook's Collage qualitatively suggest that image stills are sufficient as visual cues for jogging working memory and that the time investment of navigating and replaying video footage is not needed. Our success using image stills reinforces similar findings [2] in the work place that viewing summaries of still images was preferable to viewing video and audio replays of daily self-performed activities. We define image stills as the base unit across déjà vu displays.

8. FUTURE WORK

Future studies currently underway are further examining whether people trust the accuracy of the memory aid, and whether trust increases over time as the user gains more experience using the system. This along with the question of whether a person's reliance on the memory aid, which would effectively cause them to offload the information in their memory onto the aid, increases as a function of trust.

ACKNOWLEDGEMENTS

We thank the many psychology collaborators who refined our evaluation, especially Anne McLaughlin who inspired the dual task design. This research is funded by the Broadband Institute and Aware Home Research Initiative at Georgia Tech.

REFERENCES

1. Boreczky, J., et al. (2000) An Interactive Comic Book Presentation for Exploring Video. Proceedings of ACM CHI, 185-192.
2. Czerwinski, M., Horvitz, E. (2002). Memory for Daily Computing Events. In Faulkner, X., Finlay, J. & Detienne, F. (Eds.), People and Computers XVI, Proceedings of HCI, 230-245.
3. Eldridge, M., Lamming, M., Flynn, M. (1991) Does a Video Diary Help Recall? Technical Report EPC-1991-124.
4. Engelkamp, J. (1998) Memory for actions: Essays in Cognitive Psychology.
5. Gemmell, J., et al. (2002) MyLifeBits: Fulfilling the Memex Vision. Proceedings of ACM Multimedia. 235-238.
6. Gillie, T., Broadbent, DE, 1989. What makes Interruptions Disruptive? A study of Length, Similarity, and Complexity. Psychological Research 50, 243-250.
7. Harris, J.E. External Memory Aids. (1984) In M. M. Gruneberg, P.E. Morris, & R.N. Sykes (eds.), Practical aspects of memory. London: Academic Press. 172-179.
8. Hollan, J., Hutchins, E., and Kirsh, D. (2000). Distributed Cognition: Toward a New Foundation for Human-Computer Interaction Research. TOCHI (Vol 7, issue 2), 174-196.
9. Hudson, SE, et al. (2003). Predicting Human Interruptibility with Sensors: A Wizard of Oz Feasbility Study. Proceedings of CHI. 257-264.
10. Hutchins, E. (1995) Cognition in the Wild. MIT Press.
11. Kawamura, T., Kono, Y., and M. Kidode. (2002) Wearable Interfaces for a Video Diary: towards Memory Retrieval, Exchange, and Transportation. Proceedings of ISWC. 31-40.
12. Minneman, SL, Harrison, SR. (1993) Where Were We: Making and Using Near-Synchronous, Pre-narrative Video. Proceedings of ACM Multimedia, 207-214.
13. Shell, J., et al. (2003). eyeCOOK: A Gaze and Speech Enabled Attentive Cookbook. Video Proceedings of Ubiquitous Computing.
14. Siio, I., et al. (2004). Making Recipes in the Kitchen of the Future. Extended Abstracts of CHI 2004. 1554.
15. Tran, Q., Mynatt, E. (2003). What Was I Cooking? Towards Déjà Vu Displays of Everyday Memory. Georgia Institute of Technology Technical Report GIT-GVU-TR-03-33.
16. Tran, Q., Mynatt, E. (2002). "Cook's Collage: Two exploratory designs." Position paper for the Technologies for Families workshop at CHI 2002.
17. West, Robin L. (1992). Everyday Memory and Aging: A Diversity of Test, Tasks, and Paradigms. In West, R.L, and J.D. Sinnott (eds) Everyday Memory and Aging: Current Research and Methodology. 3-21.
18. Wilkins, A.J., Baddeley, A.D. (1984). Remembering to Recall in Everyday Life: An Approach to Absent-mindedness. In M. M. Gruneberg, P.E. Morris, & R.N. Sykes (eds.), Practical aspects of memory. London: Academic Press. 27-34

SELF-PRESENTATION ON PERSONAL HOMEPAGES

Heidi Glatzmeier, Gerald Steinhardt
Vienna University of Technology, Austria

Abstract: This paper is based on a study that investigates the self-presentation on homepages on three levels. On the first level we explore the content that was used for presenting oneself on homepages and we examine if certain "dos and don'ts" have been established in the choice of content as well as in the way of presentation. The second level does a reconstruction of the significance of content for the given material within the author's scope of meaning. On the third level the study tries to find cultural explanations for the "self-representation on personal homepages" phenomenon.

Key words: World Wide Web, Home pages, self-presentation.

1. INTRODUCTION

The economist and philosopher Georg Franck (1998) reflects on contemporary social life from a specific point of view, namely the perspective of a constant struggle for attention. Attention is no longer given freely to an individual but has to be gained by fighting for it.

Whenever people meet they present themselves to their peers. Their peers then create a profile of them based on how the person presents himself: The appearance of a person increasingly becomes more important than the person's being itself (see Haug, 1971). The more successful a person is when presenting himself, the better he can impress his peers.

A possible explanation for this feature of contemporary society may be the fading importance of societal institutions, which in the past regulated who was at what time entitled to what quantum of attention. The equal treatment and equal status of all members of society, an idea strived for since the Enlightenment and enforced today (at least on paper), promises – at least implicitly - everyone the chance of achieving anything they want (see Horkheimer & Adorno, 1996). In late modernity this development led to the process of individualisation as described by Beck, which came along with the obligation for the individual to be successful in life (see Beck, 2001).

But success and social prestige can only be achieved by somebody who manages to draw attention to himself. Attention becomes a good that is in short supply and has to be fought for using self-presentation. Franck (1998) depicts this social phenomenon through the metaphor of a pub where everyone talks at the top of their voice and every single person has to shout to be heard by the others.

2. THE INTERNET – A NEW MEDIUM IS DISCOVERED FOR SELF-PRESENTATION

Familiarity with the PC and the internet is expected of people in professional life. The acquaintance with such skills promotes the use of the new media in the private domain where computers and the internet have become more and more common.

In 1996 only 4% of Austrians aged 14 and above had access to the internet from their home PCs. In the first quarter of 2004 54% of Austrians belonging to that age-group were able to access the internet from home (http://mediaresearch.orf.at). When the new medium internet made its appearance in private households private people were given the opportunity of an active-productive utilization of the new medium.[1] This pervasion of computer technology in the domestic domain brought forth some amazing phenomena. Among these are personal homepages, which primarily serve the self-presentation of their authors.

3. THEORETICAL REFLECTIONS ON SELF AND SELF-PRESENTATION

Mead describes the development of self as a process located within social interaction (see Mead 1980a, 1980b, 1983, Blumer 1978). Individuals cannot exist severed from others; and the self (or identity) emerges from social processes. Therefore the self of the single individual is not given by birth, but unfolds little by little and is formed as a result of social experience. The process of developing one's self is seen as a process of socialisation. The basis for the development of self and consequently the socialisation according to Mead is the capability of man to shift perspectives. Man therefore is intellectually capable of making himself the object of contemplation from the perspective of others and so to anticipate courses of action and their consequences. The capability of assessing consequences of actions enables us to have a presentiment about the impression of our own person on others in the course of the socialisation process. Moreover this capability is the basis for the development of strategies by which we endeavour to create the desired impression within our social environment. For the very aspect of the self that corresponds with the internalised "generalized other" Mead introduces the term "me". Thus the "me" enables us to see ourselves in the way that we think others see us. It also includes expectations of which we assume that others want us to meet. On one hand the "me" provides the pattern for self-presentation in the sense of a fulfilment of expectations and at the same time functions as a sort of authoritative control in regard to the presumed outcome of our presentation, allowing us a glimpse on our own person from the outside[2].

According to Goffman (2002) we are trying to make the ideal that we hold of our own person come true. We do so by playing various roles and by requesting the observers to take the impression that we are trying to evoke seriously. The audience is asked to believe that the character which is presented actually possesses the attributes that he seems to possess. With reference to the presented paper this means that the author of a homepage who presents himself as cool and successful asks the visitor of his page to take for granted that the presentation matches reality. Goffman maintains that the extent to which the actor himself believes in the genuineness of the role he plays differs greatly. The spectrum extends from actors who are completely taken in by the conception they have formed of themselves and are truly convinced that the impression they created is reality, to actors who do not think their own roles are true at all. So the single individual may be deceived by his own performance or regard it in a cynical way. The cynical actor does not necessarily act for his own personal gain. More often than not he deludes his audience for their own good. Goffman uses the example of a doctor who prescribes his patients placebos because they show such a heartfelt demand for them in this context. R.E. Park (1950, quoted by Goffman, 2002) maintains that everyone is always and everywhere more or less consciously playing a role and presenting the self he would like to be. That statement also applies to personal homepages that often seem to be the manifestation of desires and ambitions. The authors of the homepages we investigated, for instance present themselves together with a car (Ferrari) they would like to possess, or by photographs taken during a trip to a large city and in a comment utter their wish for a life in a major city.

In personal homepages content that is positively valued in society is used for self-presentation. This includes items such as family (spouses, kids), friends, place of residence, the own house, pets, hobbies – often hiking, biking, tennis or soccer – and also holiday trips or events such as the wedding of the author, birth of a child or a birthday party. This tendency of authors to foster an idealized impression on their audience by trying to personalize officially accepted values of society is also described by Goffman (2002).

Another aspect of self-presentation that can often be found in personal homepages and which Goffman mentions as well is systematic modesty. Goffman describes the example of second-hand dealers who try to evoke a favourable impression for themselves by conveying that second-hand articles are without value and those dealing in them are poor and pitiable. As we found out modesty in personal homepages mostly refers to the skill of the author as designer of his page. But what is the advantage of the author resulting from showing modesty? He possibly wants to tell the audience that designing the homepage in question was not a big affair and hardly challenged his knowledge of programming. But he could also be trying to tell the audience that it was not possible for him to show his real skills for various reasons – perhaps restrictions of the provider. In both cases his modesty suggests "I am able to do better than what you get to see here". Another possible explanation for this modesty presumably lies in the fact that the author actually does not consider his homepage fully perfect and he takes a critical stance towards the outcome of his work. Thus he avoids criticism by individuals who might produce a better version. Systematic use of modesty can also be seen as an effort to provoke positive entries in the guest-book of the homepage. In this case the function of the understatement may be seen as "fishing for compliments". In all three variants of explanation the reason for the marked modesty could be fear of negative criticism in a field that is of eminent importance to the author's self-esteem – expertise on the PC. Negative criticism in a field that is essential for the self-esteem implies a massive threat and evokes anxiety. A possibility to ward off this anxiety and to avoid an impending criticism is to demonstrate modesty.

4. QUESTIONS

Our paper is based on a study which investigates the self-presentation on homepages on three levels:

On the **first level** we explore the content that was used for presenting oneself on homepages and we examine if certain "dos and don'ts" have been established in the choice of content as well as in the way of presentation.

The **second level** does a reconstruction of the significance of content for the given material within the author's scope of meaning.
On the **third level** the study tries to find cultural explanations for the "self-representation on personal homepages" phenomenon.

5. THE METHOD

The methodical approach orientates by hermeneutics used in social studies (Leithäuser and Volmerg, 1988) and is supported by semiotic analysis (Eco 1972, Kritzmöller 2002) .

The material used in this analysis is a set of 60 homepages created by people for personal use and published by wide-spread Austrian providers. The selection of the homepages used was made at random.

The challenge of the study results from the non-uniform structure of the material analysed. Personal homepages are complex patterns of pictures, wording, and context as well as acoustic signals and animated pictures. Personal homepages do not follow any linear course but are connected in a multitude of ways through hyperlinks.

6. ANALYSIS OF ONE HOMEPAGE[3]

The following in-depth-analysis of a homepage should on one hand illustrate the method of working and on the other it should make clear what a homepage reveals about the author and the attitudes that he articulates.

6.1 Description

The author of the homepage is a 36-year old male. On February 1, 2004 the homepage consisted of the topics "Wedding", "Me & My Kids", "The Woman in My Life"", "The House", "Neighbour's Cats", "Santorin", "Domrep", "Photographs", "Links", "Cartoons", "Poems", "Old and New Guestbook", "Mail", and "Home" together with a link to the "Francesco Cardeloni Travesty-Homepage". By June 29, 2004 two new topics were added, namely "Past" and "Cats" and the link to the "Francesco Cardeloni Travesty" was removed.

6.2 Evaluation

Self-presentation on personal homepages as a play between desire and reality – the homepage as a "magic" place for realizing desires

The author welcomes the visitor of his homepage by quoting the following poem by Yvonne Berlanga-Navarro:

> "Future. Dream, whatever dream you have. Go, where you wish to go. Be, who you desire to be. Dare what you have always wanted to dare. Because: Someone who continues to do what he always has done, will get what he has always got." (Original in German)

By placing the poem at the beginning of his homepage – at the most prominent point – the author makes the lines *„Dream, whatever dream you have. Go, where you wish to go. Be, who you desire to be. Dare what you always have wanted to dare."* the motto of the pages designed by him and opens a play of self-presentation between desire and reality.

According to Goffman (2002) the realisation of desires referring to one's own person is effected by playing roles. That is to say we more or less conscientiously play roles, with which we are trying to be in accordance with the respective ideal, and we are anxious to make our social environment believe in the created impression, to make them think it is real. If you consider the homepage as a new, additional possibility for self-presentation we may also assume that the play of self-presentation between desire and reality, which Goffman (2002) describes, is resumed in personal homepages. An important function of the homepage could therefore consist in its being a quasi magic place that makes desires placed in it come true. This hope that is tied to a personal homepage is expressed in the creation of subtopics such as "Happiness with a new partner". "Happiness in a new relationship" is explicitly made a topic in the rubrics *"Wedding"* (wedding announcement, wedding invitation, as well as a hundred photographs of the wedding celebration), *"Dominican Republic"* (a documentation of the first journey together), in the sub-topic *"Me"* of the topic *"Me and My Kids"* with statements such as *"I have met a very special person. Everything I ever missed or dreamed of until today, has come true"* and in the topic *"The Woman in My Life"* with statements such as *"She* (note: the new wife) *simply has everything I want of a woman ...", "You are in every single breath"*, and *"I love you"*. Indirectly you can also find the new happiness in the topics *"The House"* (the author and his new spouse had a pre-fabricated house built), *"Neighbour's Cats"* (the couple will be given a kitten by the neighbour), and *"Santorin"* (description of a holiday spent together). In seven out of eleven topics the author is explicitly or implicitly occupied with the topic "new happiness through new relationship". The wide extent to which the author's mind is occupied with this topic can be seen as a reference to the importance of the new relationship for the author.

But why does the author discuss this topic, which is probably important to him in such detail on his homepage? A possible explanation could be the presumed function of the homepages as a "magic" place for the fulfilment of wishes. The repeated placing of wishes on the homepage could be seen as a sort of magic charm, or incantation, which is to secure the realization of the manifested wishes. Following the motto "The more often something is recorded on the homepage and thus mentioned before witnesses as well as shown in pictures, the higher the chance of it becoming true."

Intimate information published on the homepage and primarily addressed to the partner such as *"You are in every breath ... "*, and *"I love you"* have an embarrassing effect on the onlooker because they are not meant for him. What function might these public assertions bear? Possible explanations can be found in associations these intimate information has for the observer such as "the whole world has got to know how happy I am" so as "to make something public as evidence" or "to declare oneself for somebody in public". One aspect in the public declaration of love may be that of a demonstrative declaration for the new wife by proving that he stands by her – showing the declarations on the homepage to as many witnesses as possible. The public declarations of love could on the other hand also be seen as an attempt to clean out personal uncertainties with regard to the new partnership in order to gain certainty in respect to the new relationship. At the same time the public declarations of his love might be understood as precautionary measure in case the relationship fails; precautionary measure to the effect that if the relationship fails the author cannot be blamed for it because he evidently loves his new partner.

The author seems to go even further by creating a combined topic called "Me & My Kids". He has another wish he would like to become reality: a link between his kids and himself. He creates this link by presenting himself and his children as belonging together designing his homepage this way and so making the dream "we belong together" come true. This combined topic stands out if compared to the others. There is no other topic on the homepage that is dedicated to more than one subject at the same time. There is a topic explicitly devoted to the new wife and also particular topics for subjects such as the new house or the neighbour's cats. The fact that the only topic that is devoted to more than one subject, one with information about the author and his children, brings forth associations in the sense of "(wanting to) demonstrate togetherness" or "(wanting to) display unity".

The homepage as medial platform for the realization of wishes is again evident in the explicit self-presentation when the author shows a picture of himself standing beside his favourite car, a Ferrari. He adds, *"To my regret this car is not mine. Only borrowed for one day. Too bad! Was a wonderful experience all the same!"*

6.3 The explicit self-presentation – male, sportive, good-looking, successful

The author presents himself on three photographs:

On the first picture he is leaning against the mudguard of a red Ferrari with one leg nonchalantly propped up on the tyre. He is dressed in cowboy-boots, black trousers and a linden-green shirt, his hair is carefully styled. The scene reminds one of film shootings starring James Dean and certain associations emerge in one's mind such as "I am an attractive guy with a fast, expensive car", "I am handsome and I am good company", "women's hearts go out to me".

The second picture – a portrait – shows the author with a tanned face, dark sunglasses and a white T-shirt. He is smiling with his arms folded across his chest. The man on the photograph signals to the onlooker the impression that he is satisfied with himself and aware of his attractiveness.

On the third picture the author appears tanned again, this time lying on the beach dressed in bathing shorts. By means of this photograph his sportive looks are shown extremely well.

According to Goffman's theory (2002), which implies that we are trying to realize the ideal image of ourselves in self-presentation, the selection of the presented photos indicates that physical attractiveness is of great importance to the author and that he is entirely satisfied with his looks.

The choice of the first photo, on which he presents himself with – according to his own words – a borrowed red Ferrari shows that he wants to be seen as a man with typical male dreams like fast and expensive cars. This assumption is strengthened when considering his usage of the technical term "year of manufacture" from the automobile sector when stating his year of birth. Finally he lays emphasis on the fact that he works as an engineer for a well-known enterprise, which is still considered to be a purely male job, and shows he identifies himself with a stereotypical male image.

6.4 Self-presentation as expression of success orientation

Several hints – above all in the sub-topic *"How It All Began – An Excursion Into Cedric's* (note: synonym for the author) *Past"* of the topic *"Me & My Kids"* make one believe that success takes up a central role in the author's system of values. He comments the photograph, which shows an approximately 5 year old boy sitting in the aircraft of a children's merry-go-round, with the words *"He has always had high ambitions (though these were not always met)"*. The photograph dates from his childhood days but the comment was written by the author only recently. The statement can therefore be interpreted as an indication of the fact that the idea of having high ambitions still bears validity today. Furthermore he shows himself as a sportive boy on three out of eight photographs – tobogganing, skiing, and finally at a victory ceremony after a competition, where he apparently placed first, second or third. The comment that goes with the pictures reads, *"Funnily enough I won most of the time"* and with that he stresses the fact that he was good which again can be seen as a hint to the central importance of success in his way of thinking. The note "*which wasn't too difficult with a bunch of Dutch who weren't trained for mountain ranges"* puts his success in perspective. But it seems that he could not have this curtailment on his homepage because in the following sentence he states that his competitors had been taller than him which makes his success respectable again (*"I was standing on a pedestal – damn it! How tall was this guy?"*).

The significance of success for the author's self-image manifests itself in the explicit self-description where he mentions pool and bowling under *"all that is fun"*. With "pool" and "bowling" – he once more names activities whose gist is to match one's own skills with the skills of competitors and therein shows his striving for still more success.

The high rating of success in his thinking is even visible on the holiday photographs in a weakened form. This can be considered a particularly strong indication for the significance of success in the life of the author because normally people on holiday can indulge in idling away their time in a socially legitimized way. The author shows that even in exceptional situations like that he finds it hard to free himself from success orientation. He comments a photograph in the section *"Dominican Republic"* as follows: *"Here is a waterfall with about 18 steps. We climbed 7 of them *sweat*. What an adventurous canyon climbing!"* A second picture on which he apparently sluggishly lies on the slanting trunk of a palm tree he comments with *"It wasn't that easy *grin*"*. Both statements show that it seems to be vital for the image he has of himself and his self-esteem that he is successful even when on holiday.

The pieces of information in the *"Santorin"* topic where he states what a successful holiday maker has to do in his opinion is a similar case, namely *"... to find out quickly what one* ***ought to*** *see."*, *"A stroll along the rim of the crater is also* ***a must****"*, *"At least once you* ***should*** *get your feet burned on the black beach of lava on Santorin"*, or *"Finally I would like to say that it is* ***a necessity*** *to have seen Santorin."*

The appearance of the topic "performance and success" in various sections of the homepage can be seen as a reference to the great importance of success-orientation in the ideal self-image of the author (see Goffman, 2002).

6.5 Self-presentation as the striving for social appreciation

As described above success seems to be of central importance to the author's self-concept which can also be seen as an expression of his striving for social appreciation. The strong need for appreciation by other people is also shown in his endeavour to combine the self-presentation as being successful with the message "I am just an ordinary person and one of you". This also becomes apparent from the set of photographs entitled *"How it all began – an excursion into Cedric's* (note: synonym for the author) *past"*. The four photographs, as described above, indicating a strong success orientation are followed by three pictures showing him as an ordinary boy and juvenile. On one photo he can be seen as a school-child playing in the sand on the beach. His comment reads as follows *"Summer, sun, beach – realised very early as being pleasant!"* On the second picture the author is depicted as a juvenile with a bottle of ketchup in hand and he writes ironically *"My most important edible at that time! (Fortunately though there were quite voluminous sizes available ;-)"*. Finally he shows a photograph that was taken during everyday school-life and explains *"Several arguments with a teacher prolonged the duration of my attendance at the technical high-school considerably ;-)"* The joint statement of the three pictures could be as follows "I like what you like.", "I like to eat what you like to eat", "I had the same problems you had" and thus "I am one of you". It can also be seen as expression of the wish "I would like to be one of you" (see Goffman, 2002).

The endeavour to demonstrate togetherness is again shown in his listing of hobbies. The author mentions *"my children, house and garden"* as hobbies. These are domains in which he can be sure to meet with broad social approval (see Goffman, 2002).

Another clue to the presumptive endeavour to be seen as a sociable person can be found in listing billiards and bowling as hobbies. Both games are played together with other individuals. The information that the author likes to play both of these games can at the same time be understood as information that he often likes to spend his spare time in the company of others.

A further central aspect in the self-presentation seems to be the attempt to counteract the image of the dull and serious engineer. Thus the author emphasizes that his favourite spare time activity is *"everything that is fun"* and then mentions billiards and bowling. As hobbies are activities that are pursued out of one's own free will one can assume that these activities are pursued for fun. Why then does the author explicitly mention them as of being fun? One possible explanation might be the wish to state that he is by no means a dull engineer but that he has a sense of humour and wants to enjoy himself. The desire to be thought of as somebody who you can have fun with makes one assume that the author thinks it as socially desirable to be taken for a nonchalant and funny person and he wants to meet what is supposedly expected of him by designing his homepage.

The endeavour for the image of an easy-going person is made clear by the cartoons he placed on his page. Five of the six cartoons deal with the topic engineering in an ironic way. Three cartoons are devoted to the topic *"MIR"* one to the *"Mars-landing"* and one to *"Sex in the internet"*. This ironic exposition of the engineer with the topic engineering makes him seem to be a representative of his group who is capable of laughing about jokes that are about topics important to him.

6.5.1 Holidays, nature photographs, IKEA, and Ö3 – tokens of unconventionality within the social standards and a standardized escape from everyday life in the success oriented society

The author dedicates two out of a total of eleven topics on his homepage to the presentation of holiday trips so that one can assume that holidays rank highly to him. The two holiday trips described are about a fortnight's vacation in the Dominican Republic and a fortnight's stay on the isle of Santorin. Currently the vacation in the Dominican Republic is illustrated using eight pictures, but a message states that this topic will be expanded in future. The pictures bring up associations of a "vacation in paradise". They show a deserted beach with palm trees and a waterfall – where no tourists can be seen, the author on the deserted beach, on the slanting trunk of a palm tree and the author standing on his own in the high waves.

In the "Santorin" section comments such as *"You notice how you get calmer, more relaxed and the problems spinning around in your head fade away ..."* indicate the assumed importance of the holiday trips as a sort of standardized frame for a temporary escape from the success orientated society.

In the *"photographs"* section the author presents the beauty of the landscape and nature. There are photographs that he shot himself as well as 12 photographs he downloaded from the internet. In addition to the pictures he offers explanatory information such as *"my photograph of the comet"*, *"a view from the Koralpe in Carinthia"*, *"The Koralpe again, higher up, some years later"*, *"The waterfall on the Koralpe"*, *"Have you ever seen Vienna in the morning?"* The selection of photographs and the subtitles clearly show that the author perceives the beauty of the landscape and of nature and that he enjoys it. The following comment on a photograph that depicts a snow covered forest in the orange morning light shows that love of nature particularly well: *"My absolute favourite"* he writes. Photographs displaying the beauty of nature for him could be a further possibility of temporarily escaping the above mentioned success orientation that is so important to him; a possibility that involves only little time and financial expenditure.

The information that the journey to the Dominican Republic had been offered in the IKEA-family catalogue can bee seen as indication of a striving for non-conventionality within the social rules and standards: On the one hand the trip was not booked at a travel-office like people usually do; on the other hand the positioning of IKEA on the market expresses the enterprise's philosophy of being different to other enterprises. Well-known slogans of IKEA are for example "The furniture-shop from Sweden that is a bit different to others" or "The unusual furnishing-shop from Sweden". In its line of advertising IKEA carefully maintains the image of a young, uncomplicated enterprise by offering furniture and fixtures in modern design and of good quality at moderate prices and at the same time emphasizes its social commitment (improvement of working conditions and no toleration of child-labour with its suppliers) as well as pleading for the protection of the environment (co-operation with WWF).

The two photographs published in the *"photographs"* section were taken on a visit to the Ö3 (note: Austrian broadcasting station) studios and can again be understood as striving for non-conventionality within the social rules and standards. The author comments these pictures as follows: *"Kudos to THAT Ö3-team! Continue your great work!!! (unfortunately they didn't manage to maintain that quality) Thank you L1SA!!"* Similar to what IKEA stands for today, namely a non-conventional life-style, Ö3, in the early years of the author (year of birth 1968), was the broadcasting station for the young generation. Ö3 broadcasted modern and consequently mostly English songs as well as ideas on socio-political topics which were considered popular and modern. Disappointment with the present style of the broadcasting station, proved by statements such as *"Continue with your great work!!!! (unfortunately they didn't manage to maintain that quality)* and *"Unfortunately Ö3 wasn't able to think of anything new. What a pity ……"*, could be attributed to the fact that in the last years Ö3 has had to compete with several privately run stations of similar styles and therefore is not as outstanding as it used to be and has lost its uniqueness. Also, individuals of the author's age aren't the primary target group of the radio station and therefore do not feel as attracted to it as they did when they were younger.

The detailed presentation of the two holiday-trips, the nature photographs as well as the references to IKEA and Ö3 can be seen – in the sense of a self-presentation as realized ideal image of one's own person – as expression of the author's wish to be thought of as unconventional and to escape from success orientated everyday life without disappointing the alleged expectations of society.

6.5.2 Traditional understanding of gender roles and passed-on rituals in the digital age as positioning of the private individual within the extended traditional frame

In the "*The Woman in My Life*" section the author depicts an image of women that corresponds to the traditional view of the role a woman has to play in our society. By stating that his wife has everything that he wants of a woman and by listing up the appraised characteristic features he at the same time informs the onlooker of his ideal image of a woman. The author tells his audience that his ideal woman must have characteristics such as kind-heartedness, tolerance, and beauty. She should always be good-humoured as well as be a good mother and a perfect entertainer. By wanting these characteristics he creates and validates an image of a woman that corresponds to the image of women within the traditional understanding of gender roles. The expressed desire for a partner who feels obliged to the traditional gender-role makes us assume that his own way of seeing himself and consequently the ideal image of himself corresponds to the traditional male gender-role. According to this appreciation of the gender-roles the tasks of a woman consist of supporting her husband, so that he can be successful in his professional life, and to take care of the education of the children. Furthermore the duties of a wife are to be attractive for her husband, to be a good companion, and to be a good hostess.

Moreover it is striking that the "*Wedding*" section is placed first on the table of contents of the homepage and that much attention was paid to the layout and design of this section. This is an indication of the great significance the wedding has had in the life of the author and it shows the high symbolic value of ceremonies for the author. An additional indication of the importance of the wedding can be seen in the supplementary sub-section "*photographs*" (note: of the wedding) in which the author publishes a hundred photographs of the wedding-celebration. The wedding-ceremony at the registry office depicted on the photographs (the exchange of wedding rings, the signing of the wedding book), the celebration following the wedding (cutting of the wedding-cake together, barriers built-up in front of the entrance-door to the house of the couple), as well as the clothes of the bride (long, white wedding-dress, white pearls in her done-up hair, bridal bouquet primarily consisting of white roses), all speak for a choice of a traditional frame for the celebration.

The design of the *"Wedding"* section – in regard to its contents – reminds one of a classical conventional postal wedding announcement sent by parents of an engaged couple or by the engaged couple themselves. An old tradition has found its way into a new medium. This old custom of sending out wedding invitations still seems to be an apt form of organizing the celebration but there are some major differences between a web page and the conventional form of postal delivery: By placing the message on the homepage it is not exclusively addressed to involved individuals but reaches an unlimited number of people. Many of them have no personal relationship to the engaged couple. Designing a unique invitation quickly has become feasible by using computer programmes. The author of the homepage in question did the following: he placed an extract from the story *"Do you know how much I love you"* by Sam McBratney at the front of the page and uses the illustrations by Anita Jeram as a graphic frame of the section. The seeming uniqueness and personal note of this computer designed wedding announcement is, when looked at closely, a skilful combination of pre-fabricated stencils.

6.6 Itinerary – self-presentation as staged expertise

Currently the holidays in the Dominican Republic are illustrated by eight photographs. On five of them one can see either the author or his companion. This gives the compilation a very personal touch.

The selection of pictures the author chooses for the "*Santorin*" section is made using completely different criteria. In the whole compilation of 17 photographs not one can be found that shows the author or his partner. The photographs reveal – similar to professionally made travel-guides – the sights and natural beauty of the country. There is for example "a view of the Caldera", "a view of the crater-island of Nea Kameni", "the windmill at Oia", the "Sunset at Firá", the "Castell d'Aquila" or "an outlook on Perissa". The photos are commented and provide useful further information in the way that a travel guide would do this. In addition to that the author furnishes the reader with practical advice about good beaches, best time for travelling, climate, means of transport and good food and drink. He words the above-mentioned tips as follows: *"In late September it is still rather hot but not unbearably so, you have got the beach all to yourself and even the water is still warm. For the duration of our stay we hired a scooter. Costs are tolerable and the island is so small that you can drive round it within one hour. Only the wind causes problems. It blows constantly, which makes travelling by scooter rather dangerous at times. I need not mention details about how people in Greece drive. There is not much to say about food and drink. If you stay away from tourist centres you will get good food."* In his résumé the author once more expresses his wish to be seen as an expert and adviser with the words *"Maybe my report will help someone* (note: the visitor of the homepage*) with the decision of where to go on his next holiday"*.

The design of the itinerary can be seen – in the sense of self-presentation as realized ideal (see Goffman, 2002) – as expression of the great importance for the author to be accepted as an expert.

6.7 Presence on the Internet raises the significance of one's own life

The author introduces his homepage with the words *"These pages are totally absurd to some extent ..."*. Why does he take all the effort to design a web page whose contents he declares to be absurd? One answer to this question could be: Because it is important to him to be present on the internet with a personal homepage. One indication of the confirmation of this assumption is the choice of content for his homepage: The author expresses wishes, introduces the people important to him (himself, children, his new partner) and shows the things (house) and events (wedding, holidays) that are of importance to him. All this produces the impression that the author, by showing on his page the people, objects and events that are important to him, wants to bestow even more importance upon them. It seems that the author wants to transpose the wide-spread opinion that "all important things are in the newspaper" on to his homepage, believing that anything that goes public on a homepage gets more weight due to its publication.

7. CONCLUSION

Summing up one can say that the author uses the homepage as an additional possibility of self-presentation. By doing so the homepage becomes a quasi "magic place", that could contribute to the realization of wishes and desires. Regarding this he uses two strategies: On the one hand he tries to make his wishes come true by constantly repeating them in front of witnesses; on the other hand he directly makes his wishes come true through creative transformation on his homepage.

Furthermore the author communicates his self-image and expresses the norms and values that he thinks are important and valid: success, traditional gender roles, non-conventionality within social rules and standards. These consequently show his view of the world and at the same time the author transports important aspects of the ideal image of himself (see Goffman,

2002, Mead, 1980a). Moreover he shows how high social appreciation ranks in his thinking.

Using the selection of content presented on the homepage he shows that people, objects and events that are important to him find their place on his homepage, and therefore he wants these to obtain additional importance.

The presented homepage is a typical example of a personal homepage. The homepages analysed by us do differ by various criteria, but the elements and dimensions that we found in the homepage presented in detail can also be identified in the other homepages that we looked into.

The analysis of all the personal homepages examined by us until now taken as a background, several generalising conclusions can be drawn from the presented in-depth analysis of Cedric's homepage:.

7.1.1 Concerning the choice of contents

The opportunity of designing personal homepages for self-presentation has become something accepted. Some homepages are organized very elaborately (for example 100 photographs of the wedding, lexical listing of private indoor plants and outdoor plants, illustration of private hiking- and biking-tours that are similar to official travel guides).

Certain topics have established themselves when creating personal homepages. Among them are self-descriptions, childhood memories and memories of one's youth, people close to the author (partner, children, parents, friends), pets (primarily cats and dogs), houses and (not as common) flats, place of residence, the car and/or motorbike, holiday trips, free time activities (above all sports and creative activities) and information on important events in the life of the author (wedding, birth of a child, birthdays).

Besides being used for self-presentation in the narrow sense personal homepages are also used as medium for the publication of works in the scientific field written by the author of the homepage such as project papers written in the final school years, diploma thesis, thesis, but also self created art such as poems and guides. The personal homepage enables the author to publish his own work independently of publishers and so enables him to keep his autonomy.

7.1.2 Concerning the immanent significance of contents within the author's scope of meaning

Very often wishes and desires are voiced on personal homepages. The creative design of the homepage makes it possible to accomplish, at least partly, one's own wishes (e.g. photograph with one's favourite car). By creating the homepage (above all in sections named "fun" or similar) personal fantasies are made public.

Objects and events are presented on the homepage that have (mostly positively) impressed the author. Also occurrences that fascinate authors are presented on homepages (e.g. a young author living in a small town reveals his fascination for high buildings and tells how much a trip to a big city had impressed him).

The wide-spread installation of guest books can be seen as expression of a hope for a positive feed-back and thus as a strengthening of self-esteem. Counters carry a similar function. The higher the number of visitors the more the self-esteem is given. In the self-presentation the author reveals the norms, values and attitudes that are important to him (e.g. the importance of success, traditional view of women's role in society).

7.1.3 Concerning socio-cultural aspects

Personal homepages can be seen as an indicator for the disappearance of the dividing line between private and public life in society (see Steinhardt 1999).

Personal homepages also indicate a high rating of the importance of a media presence in society. Prestige in contemporary society seems to be linked with media presence. Presence in the internet through personal homepages can therefore serve as an attestation of one's personal importance. In modern society access to the media and thus the possibility for spreading information in public was increasingly restricted to a few people only (Habermas 1990). Because of their ease of access homepages can now serve as a platform of speech for many. For quite a number of people this raises the question what the content of these public statements is. As described above very often one's own person and one's own life are the subject of a personal homepage. The act of articulation seems to be more important than the actual information (see Steinhardt 2002). But if so many people articulate themselves in the internet: who is there left to read the enormous number of publications that flood the internet? Does this development lead to white noise in which the act of communication is drowned out? If this is the case, G. Franck would, in the end, be right with his metaphor of the beer pub where everyone talks at the top of their voice and every single person has to shout in order to be heard.

NOTES

[1] The interrelationship and the interactions between technical artefacts and people, who use them, can be conceived as socio-technical figurations. This

term refers to N. Elias' concept of figuration (Elias 1970) and stresses the network of connections and interdependences between technology and social life.

[2] The "I", however, is –according to Mead (1980a) – the moment of spontaneity and creativity against the expectations from outside.

[3] The analysis of this homepage is an extract from a forthcoming diploma-thesis by H. Glatzmeier

REFERENCES

Beck, U. (2001): Risikogesellschaft: Auf dem Weg in eine andere Moderne. (16. Auflage). Frankfurt am Main (Suhrkamp)

Blumer H. (1978): Der methodologische Standpunkt des symbolischen Interaktionismus. In: Arbeitsgruppe Bielefelder Soziologen (eds.): Alltagswissen, Interaktion und gesellschaftliche Wirklichkeit. Vol. 1.Reinbek (Rowohlt), p. 80-146.

Eco U. (1972): Einführung in die Semiotik. München (Fink)

Elias N. (1970): What is Sociology? London (Hutchinson)

Franck, G. (1998): Ökonomie der Aufmerksamkeit. München (Hanser)

Goffman, E. (2002): Wir alle spielen Theater. Die Selbstdarstellung im Alltag. 10. Auflage. München (Piper)

Haug W.F. (1971): Kritik der Warenästhetik. Frankfurt am Main: Suhrkamp.

Habermas J. (1990): Strukturwandel der Öffentlichkeit. Frankfurt/Main (Suhrkamp)

Horkheimer, M. & Adorno, T.W. (1996): Dialektik der Aufklärung. Philosophische Fragmente. (3. Auflage). Frankfurt am Main (Suhrkamp)

Kritzmöller M. (2002): Positionierung im Lebensstil-Labyrinth. Frankfurt (Peter Lang)

Leithäuser, T. & Volmerg, B. (1988): Psychoanalyse in der Sozialforschung. Eine Einführung am Beispiel einer Sozialpsychologie der Arbeit. Opladen (Westdeutscher Verlag)

Mead G.H. (1980a): Geist, Identität und Gesellschaft. Frankfurt (Suhrkamp)

Mead G.H. (1980b): Gesammelte Aufsätze .Vol. 1. Frankfurt (Suhrkamp)

Mead G.H. (1983): Gesammelte Aufsätze. Vol. 2. Frankfurt (Suhrkamp)

Steinhardt G. (1999):Auf dem Weg zur kaleidoskopischen Wahrnehmung. Überlegungen zur Subjekt-Konstitution und Welt-Erfahrung im Zeitalter der neuen Informations- und Kommunikationstechnologien. Psychosozial 22 (1999) 1, p. 81-98.

Steinhardt G. (2002): Das Subjekt im Netz. Identität und Kommunikation im Zeitalter des Internet. psychosozial 25 (2002) 3, 27-46.

INTERNET

http://mediaresearch.orf.at

DOMESTIC TECHNOLOGIES OF THE FUTURE

Aproaching through visual scenarios

Anne Soronen and Kristo Kuusela

Anne Soronen, Hypermedia Laboratory, University of Tampere, Finland. Kristo Kuusela, University of Art and Design, Helsinki, Finland

Abstract: An overall goal of the *Morphome* project is to create design principles for domestic applications of proactive computing in a way that could support existing values of domestic life. During the first year of the project we studied meanings and understandings associated with domestic technologies and their roles in everyday life by applying the probes approach. The focus was on the material environment of the home as well as the social: important things and practices affecting cosiness, domestication of media technologies, and use routines of different kind of domestic devices. One purpose of the project is also to explore how emphatic design methods can be applied to produce understanding of people's emotional relation to their domestic environment and dynamic use contexts of domestic media and technologies. The starting point was that proactive solutions applied in the homes have to meet high standards in aesthetic and social usability in order to become widely adopted and accepted by the people.

Key words: Domestic technology, future home, scenarios.

1. BACKGROUND

The overall goal of the *Morphome* project is to create design principles for domestic applications of proactive computing in a way that takes sensitivity of the home as a use environment into account. The idea is to increase the understanding of people's emotional relation to their domestic environment and the dynamic use contexts of domestic media and technologies. During the first year of the project in 2003 a probes approach was applied in order to explore meanings, practices and values associated with domestic technologies and their roles in everyday life (see Soronen & Sotamaa 2004).

The goal was to produce a diversified notion of current domestic life in the media-rich Finnish homes before ideating product concepts that would suit the principles of proactive computing.

During the second year of the project in 2004 we have come closer to the field of proactive domestic technology. The work done in our project can be called design-oriented research in which production of new knowledge is the main contribution instead of artefacts. However, the knowledge acquired would not have been attainable if design (in the form of experience prototypes and scenarios) had not been a central part of the research process. (cf. Fallman 2003.) When it comes to research methods, our starting point can be summarised with the notion by Botero Cabrera & al. (2003) when they argue that in order for people to be co-designers discussing directions of future technology, they must be given tools and materials inspiring them to reflect on personal possibilities and alternative practices of new technology. Furthermore, we were not interested in approaching proactive computing as a tool for specific domestic tasks to be conducted as efficiently as possible but rather to consider what kind of meanings and feelings the participants would relate to the chosen application concepts from the viewpoint of their personal or familial life. An underlying idea of our work supports Hallnäs and Redström's (2001) suggestion that with the increasing ubiquity of computational things becoming a natural part of people's everyday lives we should shift focus from design for efficient use to design for meaningful presence.

This paper describes work in progress. The focus is on the goals and the implementation of the interview study in which visual scenarios were used to illustrate concepts of proactive computing. The first year's probes material functioned as a starting point and inspiration for devising scenarios. Conducting the interviews in the homes of the participants was a significant choice because the idea was that people could assess the scenarios from their own domestic context. Moreover, we wanted at least some concepts to be simple applications that could be used in existing home environments without much work of installing large-scale systems. Hence, the idea was to be sensitive to the particular homes and their practices and to give tools for the participants to think of their homes as environments in which novel proactive solutions could be potentially used together with existing ones.

Because the term 'proactive computing' is intangible and unfamiliar to most people, concretising it is a challenging task. And because we as researchers had no personal experience of testing, or living with embedded systems called 'proactive', our concept ideation was based more on what we know about contemporary domesticity in Finnish homes than what we know about proactive technology available to consumers. During the interviews we aimed at gaining a fuller understanding of people's everyday lives by focusing on their practices and activities in particular living situations instead of merely discussing our design concepts. In fact, we did not want to restrict the scenarios to represent concepts that can be categorised purely as proactive computing, since the focus was more on acceptability of novel computing technology in such an emotionally and socially sensitive use environment as the home. Hence, some of our concepts could be more accurately labelled context-aware than proactive technology. We understand context-aware technology to be a broader concept than proactive computing which is generally used to refer to technology that takes initiatives or even anticipates the needs of the user. The starting point was that proactive technologies applied in the homes have to meet high standards of aesthetic and social usability in addition to reliability in order to become widely adopted and accepted by people (see Mäyrä & Vadén 2004).

2. HOW THE STUDY WAS IMPLEMENTED

After discussing the research methods for exploring acceptability of concepts of proactive computing, we decided to organise interview studies in which scenarios would form the basis for discussion. The idea was that the first study focuses on solutions that would be fairly easy to implement now or in the near future and the second one will represent proactive technology that is more futuristic or even utopian.

The designer of our team used a lot of time to come up with new concepts and questions to be studied. After several common brainstorming sessions we were able to choose seven concepts to be used in the study. The selected concepts were related to domestic routines, comfort, security and "being together" in the home. The purpose was that the scenarios would work as flexible conversation points illustrating possibilities of context-aware technology to be used in a domestic environment. The concepts ranged from a small trash-can-like object informing one when the real trash can in the kitchen is full to an application playing a fanfare when members of the household or relatives assemble in the home for a special occasion or just as a part of everyday life. We also had a more familiar "smart home scenario" in which networked components adjust the quantity of light, temperature, and ventilation and are connected to some domestic appliances, such as a coffee maker or an alarm clock. From the viewpoint of technology the concepts ranged from simple applications to larger systems more difficult to implement.

Each scenario was composed of 2-5 pictures in which different situations of use or merely a product (involving embedded computing technology) were visually depicted. During the interview the researcher explained some alternative functionalities of the concepts while the participants glanced at the pages of the booklet in which the scenarios were put together. The booklets were used because they were easy to carry along when doing field work and because the participants could thus personally handle the visual information they were offered (cf. Keinonen 2000). The concepts were represented as drawings illustrating situations of use or functionalities of the concepts because we assumed that it would be easier for the participants to imagine use of the novel applications in their own domestic life if they were not related to any particular users. Sketchy drawings, the scenarios were more like open-ended descriptions of use situations of different applications (see Figure 1) than detailed descriptions of determined features of ready applications.

Figure 1. Two example pictures from the same scenario illustrating a possible situation of use and functionality of the concept.

Before the interview the participants were sent priming questions related to their domestic everyday lives and also to the use contexts of the concepts. The questions were sent to the participants beforehand in order to 'tune them in' to the themes of the scenarios since the topics of the questions were more or less related to the intended use of the application ideas. The scenarios also functioned as a basis for a dialogue between the researcher and the participants. On the one hand, the objective was to make concepts so accessible to the participants that they could piece together whether the proposed application would suit them or not. If the priming questions motivated people to think about the interview themes, the scenarios motivated them to produce personal descriptions about the potential or undesirable situations of use for the concepts in question. The priming questions concerning the scenario in Figure 1 were: "What kinds of sounds disturb you at home?" and "Have you ever needed ear plugs at home?"

Our scenarios differed from those scenarios in which designated users of the product, their objectives and activities with the product and the context of use are written down in a narrative form (see e.g. Carroll 1999). Although the scenarios represented users of different ages and gender they remained anonymous 'flat' characters whose role was minor in the discussions between the researcher and the participants. The appearance, clothing, ethnic identity, gender and age of characters affect the participants' thinking to some extent but during the interview attention was directed away from 'particular characters' towards the participants themselves as potential users. Hence, the scenarios were not intended to be character or plot-driven stories but application-driven storyboards in which pictures combined with the verbal explanation of the interviewer formed a starting point for conversation. Some hints of the technical possibilities were presented, but no proper use cases. They were hidden on purpose. The interviewer also had predefined themes and some detailed questions that were discussed before moving on to next scenario. The aim was that the drawings, together with the interviewer's arguments, would invite the participants to create stories about appropriate and inappropriate uses for the concepts in their own domestic contexts. The purpose was also to attain information to be used for selecting our forthcoming experience prototypes in the Morphome project. The study confirmed some of our own insights but also gave us new ideas and information for prototyping.

3. TENTATIVE RESULTS

Twelve households living in apartment buildings in Tampere and Helsinki were recruited for the study through an advertisement in a local newspaper and on the website of the Hypermedia Laboratory (www.uta.fi/hyper/). In the advertisement we highlighted that we were looking for households that are interested in discussing everyday meanings of domestic technology and assessing product concepts of the future from the viewpoint of their own homes. Hence, the participants supposedly had a more positive attitude towards technology than people on average or at least they were more interested in reflecting on aspects of new technology.

The households were selected to represent different modes of living (including a single person, families with and without children, a couple with adult children living in their own homes, households with and without pets) and diversity in ownership of media technologies. The average age of the participants was 30 years, the youngest participant was 13 and the oldest 52 years old. We had 15 people in the working life, seven students, and one pupil in an upper level of comprehensive school. They all had variety of media technologies, for instance, every household had at least one computer and radio while none of them had a landline telephone. This can be related to the prevalence of mobile phones in Finland. Among the participants we had also a family who had no television.

Next we highlight some points that will hopefully illustrate the diverse attitudes the participants had towards the concepts of proactive applications. We will not concentrate on the particular scenarios as much as on common opinions related to aspects of awareness of functionality of novel context-aware domestic technology.

3.1 Proactive domestic technology – the master of the house or a domestic help?

Awareness of the functionality of automated domestic systems is closely related to the feeling of control. Most of the suspicions the participants had towards proactive technology were linked to decrease in control over automated home systems. For western people who live in an urban area, the domestic space is very often the only territory in which they can arrange the environment as they wish. In that sense it was not surprising that the participants were afraid of losing the feeling of control when living surrounded by proactive technology that senses the environment. Most of them were afraid of that proactive systems make decisions on their behalf and a possibility of misjudgements was felt very high. The fears were especially associated with situations in which the system would "assume" the dweller's state of mind or the next activity. Most of the participants suggested that it should be possible to 'turn off' such an automated system or at least some of its functions. However, they supposed that in many cases they could accept the proactive system taking initiatives if it was just offering alternative functions to be chosen rather than making decisions on the grounds of predefined 'parameters'. Thus, the participants highlighted that they could approve such a system on the condition that they have a possibility to 'cancel' a suggested function.

Depending on the concept, the participants' reactions to the users' sufficient awareness of technology's functionality and presence varied a lot. For instance, when the focus was on the smart home scenario some of the participants considered that manually opening the curtains in the morning makes sense. The underlying idea was that this kind of a morning routine belongs to the waking up process and it is not a good thing to turn it over to a proactive system. However, most of the participants thought that they could accept a dim light switching on automatically simultaneously with the sound of an alarm clock.

The discussions about a pedestal lamp measuring the domestic decibel level and indicating the current level with changing colours also illustrates the diversity of attitudes. Some participants perceived this kind of a decibel lamp to be a furnishing element that should work silently in the periphery of their awareness. A few pointed out that the alteration of colour is too strong of an indicator of change in sound level. They preferred the lamp to inform them about changes in the decibel level just by changing the intensity of the same colour. However, a majority of the participants saw the lamp primarily as a functional object that they would locate in a visible place within their home. They thought it was a pleasant idea that the lamp could offer information about the sound level which can be difficult to assess oneself. Some participants mentioned that the dwellers' ears are prone to adjust to their own domestic soundscape and at times it is very hard to know if the sounds carry to the neighbouring dwellings.

Compared to the other participants, comments diverging to an extent were voiced by a man who considered that in their home the decibel lamp would work best if it was an unobtrusive and decentred technology situated in a bedroom, the living room, the kitchen and possibly also in other rooms. In the living room the man would have placed it either on the bookshelf or next to the television where it can be easily observed when one is watching the television. On the other hand, he suggested that information offered by the decibel lamp might as well be presented on a computer screen. He remarked that the computer in their home is always on when he or his girlfriend is awake and therefore the 'lamp icon' could show the current decibel level, e.g. flashing when the level changes. In the latter case he related easiness and non-obtrusiveness to getting the information from the computer screen. Hence, the computer as an object was already felt to be a self-evident part of the visual order of the living room. Because the computer was frequently the target of attention the participant felt that it would also be appropriate to get sound information from the computer screen. He felt a new visible and distinct object to be potentially disturbing and preferred the idea that the lamp should be fairly imperceptible but easily available. Hence, he assumed that if the decibel lamp was a visible object, the place next to the television would probably ascertain that one would not have to shift attention from the television to other directions too much. In his arguments centrality and accessibility of "screen media" in the domestic environment was unquestioned.

3.2 Is calm technology a solution?

The notions and reactions relating to the scenario of the decibel lamp can be considered together with the idea of calm technology, in which the presentation of information in a non-obtrusive way is pursued. According to Weiser and Brown (1996), calm technology moves easily from the periphery to the foreground of our attention, and back. They see this calming effect happening because people are able to attune to many more things when most of the technology is placed in the periphery of attention. In their vision information technology comes to the centre of attention only when it is needed but the users remain in control because through peripheral awareness they are attuned to the technology while simultaneously free to do something else. On the grounds of the decibel lamp example, it is really hard to define when information is sufficiently in the foreground of attention and, on the contrary, when it is sufficiently in the background of attention. For instance, the red colour indicating the most stentorian sound levels was perceived to be potentially disturbing because it requires a lot of attention and sometimes you cannot avoid loud situations in the home. In this case, the lamp would irritatingly strengthen the awareness of stentorian sound levels although you could not necessarily change the situation, e.g. if you had an ill baby bawling or a painting had to be fixed on the wall by using a drill.

The discussions with the participants indicated that the home as a use environment changes when different dwellers and especially visitors are present. With visitors people are prone to think more carefully which devices are allowed to be in the centre of attention and which ones are not. For instance, a middle-aged couple said that they always switch off the television when visitors are coming because they do not want the television to take too much attention and steer the conversation. Furthermore, many domestic appliances (such as a vacuum cleaner, a washing machine or a tumble dryer) were felt to be so loud that it is not possible to move them to the periphery of one's attention when they are in use. When asked if the participants would keep the decibel lamp on when they had visitors the answers were twofold. Some felt that it would be impolite towards the visitors if the decibel lamp would show that talk was too loud by its colour. Meanwhile, other participants argued that it would be just fun to have parties with the lamp turned on because it would be a playful focus among the visitors.

In western countries homes are perceived more and more as centres of individualized leisure culture (Livingstone 2002). At home people can choose fairly freely how they organise time and space, what activities they undertake and how they make the environment more agreeable and comfortable to themselves (Meyer & Rakotonirainy 2003). Although the participants of our study were technologically oriented people, most of them highlighted that they want to have domestic moments when most or all media technologies are switched off and during these moments also peripheral awareness of technology is felt to be irritating. Thus the idea of continuous presence of embedded proactive system was mostly questioned whereas simpler proactive applications were more easily accepted. The home was often perceived to be the 'last' place where amount of information can be delimited and maintaining of this limitation was probably easier to link to the single applications than to the large-scale systems. However, some participants had positive attitude also to the larger proactive systems in the case their functions were confined to specified tasks.

As expected, people were prone to think of new proactive technology on the grounds of their experiences with existing technology. The idea of embedded proactive technology could be felt dubious because people are used to perceive domestic technology as visible and concrete devices that they can place and control. When the technology and its functionalities transforms into something that is embedded in the domestic surroundings in pursuance of taking initiatives for "domestic action", there emerges a threat of losing the control and knowing insufficiently what the domestic technology is doing. Calm technology can sound ideal but in the contemporary media-rich homes it is not so simple to determine which information or functionalities should be in the periphery and which ones in the foreground of attention because the use of technologies seems to be fairly dynamic and situation-specific in nature.

3.3 Remote awareness of the condition of the home

If the home is typically perceived as a shelter for its dwellers, it is also seen as a place that has to be sheltered. The commonly shared idea among the participants was that after the workday, school day or a holiday trip the home is basically expected to be in the same condition that it was left in. It was not surprising that the participants were eager to know if something untoward is happening in their home when they are not present. According to Meyer and Rakotonirainy (2003) security systems that increase awareness of the condition of the home belong to the recurring visions associated with context-aware homes. Hence, the participants were presumably familiar with the idea of remote emergency alarms informing the dwellers about 'harms' in their homes. The automated alarm systems increasing security or at least the feeling of security seemed to be desirable both among home-owners and tenants because the home with its personal and memorable objects has strong emotional value. Many participants experienced their home to be some sort of an extension of themselves and the threat to domestic property was felt to be a threat to one's own subjectivity. The home's "intactness" seemed to be on the minds of the participants fairly often when they were away from home.

Most of the participants felt it to be meaningful to use mobile technology that could inform them in real-time about material damages or emergencies going on in their home, even if they did not know whether the alarm was related, e.g., to an intruder or damage caused by water or fire. This discussion was related to the scenario in which people came up with ideas about the kinds of situations in which they could receive a vibratory alarm from their home when they were somewhere else. A majority of the participants thought that an instant vibratory alarm would work well in emergency situations. Most of them wanted to receive alarms with their mobile phones but some favoured the key chain. However, a few participants thought that the mere vibratory alarm would not be a satisfactory information source and they also wanted to receive verbal information about what had actually happened if they were to use an alarm system. An exception was a female participant, a mother of a 6-year-old girl, who thought that the vibratory alarm would be most useful to her when her daughter starts going to school. For instance, the vibratory alarm could inform her when her daughter has left for school in the morning and when she is back from school. So, the mother thought that the vibratory function would be more useful to her when used to inform her about her daughter's school-day comings and goings instead of emergencies which usually occur rarely.

The desire to be made distantly aware that everything is alright at home was limited to the indoors of the dwelling. For instance, nobody mentioned that they would like to receive these "damage reports" relating to the whole block. The alarms from home received by a mobile phone were seen as a natural information channel probably because both are personal objects that people want to protect from outsiders. Just like many users understand the mobile phone as a 'body part' (Morley 2003) some of the participants talked about their homes as extensions of themselves. However, a few participants indicated that they would prefer to use the vibratory function received by a ring or a bracelet because they are always worn on their bodies and compared to the mobile phone the vibratory notification would be more easily noticed when it happened on the skin.

Awareness and control of proactive computing technology are very challenging issues especially when designing applications or systems for domestic environments. Generally speaking, people were of the opinion that they want to be highly aware both of the condition of the domestic space (especially when not present in it) and the availability of functionalities or information the technology offers. They also argued that, in most of the cases, they want to keep control and avoid systems in which technology can make decisions on their behalf. Although the participants in question were interested in new technology and its possibilities, they highlighted that they do not want to have new futile technologies in their homes. The underlying idea was that among new domestic technologies there are lot of useless devices that are not appropriate or meaningful for long-term use. In this regard, embedding new features into existing domestic products seems to be a direction that has to be taken into consideration more carefully in the near future. However, designers and researchers should also consider to what extent existing domestic objects, such as furniture, can be transformed by adding embedded computing features without "disturbing" the homely aspects and comfort of domestic environment.

4. DISCUSSION

By using the scenarios we were able to illustrate some possibilities of domestic technology of the future and they also functioned as tools for conceptualizing different approaches to it. It was important that the interviews were conducted in the homes of the participants because the feasibility of the conceptual ideas was easier to contextualize for both the participants and the researcher in the participants' homes. The scenarios without a story line or designated characters induced the participants to further develop the concepts and come up with stories of their own for possible uses. The scenarios together with the priming questions diversified our understanding of the variety of the participants' domestic practices related to the specific situations and the time of day. Nevertheless, although the interview based on scenarios seemed to work well in a concept design phase, it is not a sufficient participatory design method when the aim is to produce new products. In order to bring out more detailed demands on the features of the form, appearance and user interface of new technology, people need to have personal use experience of the product or prototype.

The next round of scenarios will concentrate more on controlling and alternative means of being aware of the surrounding proactive technology. We are especially interested in the following questions: Is it possible to 'teach' a proactive system? How do users experience the potential error conditions in different situations? What is a meaningful degree of proactivity (so that it will not affect the nature of domesticity and cosiness too much)? In addition to the scenarios we will build prototypes that participants can test in their homes to get concrete ideas of possibilities of proactive technology to be used in the home.

ACKNOWLEDGEMENTS

We would like to say warm thanks to all the participants of the study. We also want to thank the other members of the Morphome team and the Academy of Finland for the funding.

REFERENCES

Botero Cabrera, A., Kommonen, K-H, Oilinki, I. & Koskijoki, M. 2003. Co-Designing Visions, Uses, and Applications. *Proceedings of the 5th European Academy of Design Conference.*

Carroll, J. M. 1999. Five Reasons for Scenario-Based Design. *Proceedings of the 32nd Hawaii International Conference on System Sciences.*

Fallman, D. Design-oriented Human-Computer Interaction. *CHI Letters* , vol. 5, issue 1.

Hallnäs, L. & Redström, J. 2001. From use to presence: On the expressions and aesthetics of everyday computational things. In Redström, J. *Designing Everyday Computational Things*. Doctoral Dissertation. Available at http://www.math.chalmers.se/~redstrom /thesis/hi/use2 presence.pdf.

Keinonen, T. 2000. Pieniä tarinoita pienistä puhelimista. In Keinonen, T. (ed.) *Miten käytettävyys muotoillaan?* University of Art and Design Helsinki, 207-220.

Livingstone, S. 2002. *Young People and New Media. Childhood and the Changing Media Environment.* Sage Publications, London.

Meyer, S. & Rakotonirainy, A. 2003. A Survey of Research on Context-Aware Homes. The paper was presented at a Workshop on Wearable, Invisible, Context-Aware, Ambient, Pervasive and Ubiquitous Computing. *Conferences in Research and Practice in Information Technology*, vol. 21.

Morley, D. 2003. What's 'home' got to do with it? Contradictory dynamics in the domestication of technology and the dislocation of domesticity. *European Journal of Cultural Studies*, vol. 6, issue 4.

Mäyrä, F. & Vadén, T. 2004. Ethics of Living Technology. Design Principles for Proactive Home Environments. *Human IT*, 7 (2). Available at http://www.hb.se/bhs/ith/2-7/index.htm.

Soronen, A. & Sotamaa, O. 2004. "And my microwave is a fox": Reflecting domestic environments and technologies by means of self-documentation packages. *Proceedings of the Fourth International Conference on Cultural Attitudes towards Technology and Communication 2004.*

Weiser, M. & Brown, J. S. 1996. The Coming Age of Calm Technology. Available at http://www.ubiq.com/hypertext/weiser/acmfuture2endnote.htm.

ETHICAL ASPECTS OF HOME INFORMATICS AND TELEMATICS

Andy Sloane
Research Institute in Advanced Technology, School of Computing and IT, University of Wolverhampton, U.K.

Abstract: Home-oriented Informatics and Telematics (HOIT) research has a number of challenges in the future. These range from the technical challenge of providing useful solutions through the ubiquitous paradigm to the cultural challenges of the 21st century. While much research focuses on the technical processes of devices, services and systems for the home there is little reflective research that looks into the ethical components of these developments. This paper outlines some of the ethical problems that are faced by developers and researchers in this subject.

Key words: Ethics, sustainability, market forces, technology.

1. INTRODUCTION

The home is a concept that is rooted in many traditions and cultures of the world (Sloane, 2003). Different concepts of home are relevant in different cultures; and within these there are many different instances of "home". This has made generic study of "home-life" very difficult over the years for the many researchers that are interested in the subject. (Sloane, 2002).This variability in the manifestation of the home concept has led to many generalisations in the design and production of devices, systems and services for the home that do not give the optimum service for individuals. The rooting of the home in its culture is fundamental - it is not a standalone concept devoid of influence from the embedding social fabric, but an integral part of society, neighbourhood and culture. In many modern societies there is a move towards multiple co-existing cultures with a wide variety of home styles and configurations in evidence. This leads to even more disparity between the perceived usefulness of home systems design and their actual use and application. To encompass this diversity is one of the challenges of HOIT for the future, but to do so requires a more ethical approach to the development of services and systems for the home environment that will incorporate the cultural diversity in different countries, and the diversity in multi-cultural environments and foster the development of devices and system that aid and assist the user where necessary without causing more problems than they solve.

2. MARKET-LED DEVELOPMENT

The main source of devices, systems and services for home users is the industry that sells these to consumers. The marketing push in the industry is clearer to engage a wide range of users in the process of purchasing new equipment and systems for the home. Some technologies have been more successful than others and there are a number of historical studies that show that wider take-up is not necessarily because of superior technology (see Liebowitz and Margolis, 1995) and the outcome is often closer related to marketing, advertising and other non-technical factors rather than any technological, or even usability, superiority.

The market background document from the TEAHA project (Homega Research, 2004) gives some insight into the status of various technologies and their take-up around the world. This allows some perspective on the market for connected home products. As they point out in this report the prediction of markets is fraught with danger and simple extrapolation is not a reliable indicator of future trends in such diverse markets as home electronics and leisure equipment.

The market-led nature of this technology is also a factor that does not lead to the most ethical style of development for the systems in question. The primary goal of manufacturers is to sell goods, to make profits and to continue to operate. They do not necessarily have any "emotional capital" in the goods that they sell and promote. In fact there are a number of conflicting goals for manufacturers that do not assist the user. Amongst these are the problems of compatibility and standards.

2.1 Compatibility issues

There are some complex economic issues involved in the production of goods for the HOIT market, as for many other markets for electronic goods and services. Many of these issues are related to the compatibility of different components and systems for use in the home. The issue for companies and the goods they produce (i.e. do they produce them to be compatible with others or not?) rests on many factors and is too complex to discuss in this paper. However, the more significant side of this argument, from the home perspective, is summarized by the following (Economides, 1999):

> "Society reaps significant benefits from compatibility and standardization"

It is this factor and the incompatibility of much market-driven production that are the opposing forces in the HOIT arena and that can determine the outcome of technological developments.

The compatibility of devices, systems and software is not a pre-requisite for HOIT development but it has a considerable influence on the take-up and spread of technology in the home and across cultures and countries. It is also clear that the ethical development of home systems, devices and services requires them to be developed using an ethical perspective which must account for the societal need for standardization in addition to the commercial requirement for profit. One example of how this could be done, to promote better streamed multimedia is outlined in Sloane 2002, where infrastructure developments are shown to provide better quality of broadcast.

2.2 Standards

The issue of standardization is quite complex. The opposing views are that: on the one hand, standards are necessary to ensure compatibility of equipment and, on the other, that standards are a restriction to the successful development of new devices and systems. Both of these viewpoints are, to some extent true. The use of standardization has helped many systems to gain a bigger share of the market and many manufacturers are able to compete in a market that is led by standardization activities. Examples of this are: the CD audio and DVD video standards. These have led to many different devices being available to users to play the standard software.

Standardisation of media standards helps both equipment manufacturers and content producers provide users with compatible devices and useable content. Indeed, without the standardization effort the market may not reach critical mass before the device or system becomes obsolete. There have been many different proprietary media standards that have faltered without the back-up of a wide variety of content to enhance sales.

The opposite problem of standards; that of restricting innovation is only problematic when the devices that are standardized become widespread and the standard becomes technically limited compared to other competing standards. This situation is the case with video recording equipment. There is a large-scale legacy investment in VCRs around the world and they are now much more limited than the digital replacements DVD recorders. However, there is still a need to support the VCR user with software content, consumables and maintenance for some time into the future. There are some moves to end support for VCRs by ending sales but the legacy of many old devices will continue for a number of years and there are still manufacturers supplying new devices for users.

The move from video-cassette to DVD as the medium of choice for consumers is likely to be driven by a number of factors: wider DVD content provision, compatibility with other devices such as PC drives, multi-function capabilities (e.g. on-screen viewing of photos), digital input, output quality factors, price (although VCRs are currently much cheaper than DVD recorders), and media cost (DVD+RW and Video-cassette prices are roughly comparable). However, one of the most useful and attractive characteristics of using discs is the indexing and random-access capabilities of the devices. All these factors and the market push of the manufacturers is likely to see the replacement of VCRs by DVD devices in a fairly short time.

The next problem that will be faced by home IT users is then likely to be the recording of family archives of legacy content onto new media. This has been an issue in the past, albeit on a smaller scale, when the introduction of home video recorders made home use of cine film unattractive. Some content was moved onto video-cassette by specialist companies. Upgrading to digital formats will see more legacy content being moved to new media systems.

3. ETHICAL DESIGN

There have been a number of studies that have used co-operative design methods (for example, Kristensen et al, 2003) these are perhaps the most appropriate for the design of home devices as they involve the end-user in the design process, although there are problems with the application of any methods to the design of home devices.

However, there remains the problem of how the design of home systems can be made more ethical. The initial problem of how to design a home artefact needs to be replaced by an extra stage of analysis that asks the fundamental question "Should this device or system be designed and made?" and if it is, "How should it be made so that it can be used ethically?" and finally "Is it possible to make an ethical version by excluding or enhancing certain functionality?"

With the use of a comprehensive ethical analysis of the situation of the user and the device, put into the context of the home, devices can be designed to include a more ethical dimension for the type of device that is in use by people for a significant amount of their free time. The idea of an ethical dimension to home-system design will be explored in the following sections where the design of home devices, home services and the status of the user will be outlined and discussed

3.1 Home devices

Many home devices are developed to enhance the users' leisure time activities; others are developed to be labour-saving devices, some devices incorporate an element of both functions. There are a number of design considerations that are needed when developing devices for the home. These will depend on the function and use of the device under consideration. In addition, for an ethical perspective to be included in the design process an extra stage will need to be added that takes a wider perspective at different stage of the development process. Incorporating ethical aspects into the design process is not easy and many designs assume that the devices designed will be used in particular ways (Robertson, 2004). In the home context there are other factors that need consideration. Freedom to experiment and use devices in ways that are not pre-determined by design are aspects of device development that are useful to leisure users. There are also aspects of design that are ethically questionable when wider societal issues are included in the design process. For example, the production of violent computer games is a problem area where access to such software is not restricted, reportedly causing further aggression if the users frequently play the games (Bushman 2002).

There are also areas of device design that may cause ethical dilemmas in that perceived use is not necessarily in the interests of the consumer. A useful case of this to analyse is the ubiquitous use of the remote control to control devices n the home. There are many situations when a remote device is useful, and it does allow easy and simple control of television and other household devices. However, for some of the population (e.g. the morbidly obese), excessive use of the remote control compounds their inactivity and can exacerbate their health problem. Indeed, it has been stated that the general population would benefit from the removal of remote control devices to increase activity levels. Is this a case for ethical design?

3.2 Services

The provision of digital services to the home is also an area where ethics can play a part in enhancing the experience for users and allowing a better interaction with the service provider. There are now a multitude of different services aimed at the home user with many different types of interaction allowed with the various services. Much if what is sold to home dwellers is done so under the banner of leisure and entertainment. However, this is essentially a service that only caters for a small minority of viewers in that they fit the unique market research categories used to determine programme output by the various providers. This usually entails a diet of popular work that does little to enhance the television experience and does not venture into new territory or allow for much cultural diversity. As most of the revenues of the television and other service providers depends on high volume this is not surprising, but what is now possible is the expansion of audio-visual programming to include Internet delivery in some cases which would allow the more esoteric and culturally diverse offering for home users.

3.3 Who is the user?

It may seem to be a strange question to ask but who is the user in this context. The home is a place where people live—it therefore encompasses all users to a certain extent. This is a basic problem for home-based research i.e. what is a definition of "the user"?

Homes consist of many different, culturally-dependent configurations. Any one home is different from any other sometimes in quite profound ways – in its structure, operation, context, organisation and use. The ethical approach would be to incorporate this diversity in the design process. This is obviously a very difficult process, as user-groups in the home encompass all age groups, many different states of health and well-being and different needs, wants and desires—all within the same home.

3.4 The Internet

The Internet is now the medium of choice for the delivery of online services to the home. The Internet has developed from a computer network for research and the military, into a global network incorporating both commercial and non-commercial interest and a near-complete range of electronic human activity. This has led to some disquiet in certain areas. Many groups of Internet users do not want to interact with some of the material that is contained on the World Wide Web. Examples are: the case of separating pornography and young Internet users, and the promotion of violence or illegal drug use.

There are many areas of human life that are not suitable for wider promulgation, but legal restrictions have little effect on an electronic environment that spans continents and where they have little long-term effect. The Internet has allowed users to interact on a much wider scale, to enhance their leisure activity with a wider range of sources of knowledge and to pursue new interests that were not possible before the advent of the worldwide Internet. This has led to a situation where the Internet is now indispensable for many home-based users as it allows them to interact both socially and for commercial activity in a way that is becoming symptomatic of the 21st century.

The growth of the Internet is also one reason that it has become difficult to govern. Multi-national co-operation is required to enforce any effective features on the network and this is unlikely to be a feature of any future embodiment of the Internet or the WWW. Ethical development of Internet services may be left out of plans for governance but it does not need to be excluded from user-level service provision. If user demand is sufficient for value-added services to be economically provided they will become more widely available.

3.5 Future devices and services

For future devices the design process can be adapted to incorporate some of the ideas outlined so far: greater inclusivity, wider applicability, more ethical input to functions and design and more diverse approaches to service provision. These changes will not be easy or quick to implement; a radical overhaul of design practice and service provision needs to be undertaken and there may be resistance from manufacturers and consumers to some developments if the changes are not included with a collateral change in the legal framework, the social expectations of the home dweller and the societal organisation of commercial entities.

4. CONCLUSIONS

Many developments have changed the face of the home environment over the years since information and communication technology became a consumer item. The growth of information services and devices has allowed home-users to extend their leisure and the facilities they use into a virtual world. This new paradigm for home users has caused problems for many of them. These problems are partly caused by the lack of a sufficiently ethical dimension into the design of devices and services for home users. A more ethical approach would see more inclusivity, more cultural diversity and better quality of devices and services for home-based users.

The ethical approach is, however, not necessarily a simple and easy route to follow. Human nature militates against some of the more appropriate ethical developments and cross-cultural differences are also hard to combine into an approach that is inclusive. There are, therefore many barriers to overcome on the way to an ethical approach to home systems design, but there are a number of areas that can be used as a starting point. It may be too late to influence the design of existing systems but future systems can be influenced by a more ethical approach – it is still to be seen whether it will happen.

REFERENCES

Bushman, B. J., & Anderson, C. A. (2002). "Violent video games and hostile expectations: A test of the General Aggression Model". *Personality and Social Psychology Bulletin, 28,* 1679-1689.

Economides N (1999) "Competition and Vertical Integration in the Computing Industry", in Eisenach J. A. and. Lenard T M (eds.), *Competition, Innovation, and the Role of Antitrust in the Digital Marketplace,* Kluwer Academic Publishers 1999

Homega Research (2004) "The worldwide markets for the Connected home: status and trends", TEAHA background document, Available from http://www.teaha.org/project_outputs.php (Downloaded 9th November 2004)

Kristensen J F, Eriksen M A, Iversen O S, Kanstrup A M, Nielsen C and Petersen M G (2003) "Young People in Old Cars - Challenges for Cooperative Design", Electronic Proceedings of the 26th Information Systems Research Seminar in Scandinavia/IRIS26, Haikko Manor, Finland, August 9-12, 2003.

Liebowitz S J and Margolis S.E (1995), "Path Dependence, Lock-In and History "Journal of Law, Economics and Organization.

Robertson T (2004) "Ethical issues in interaction design" http://research.it.uts.edu.au /idwop/downloads/Robertson/Ethics&IT.pdf. Viewed and downloaded 10/12/2004.

Sloane A (2000) " Infrastructure issues for Internet broadcasting to home-based users", in Beardon, Munari and Rasmussen (Eds.), "Computers and Networks in the Age of Globalisation", Kluwer, Boston, ISBN 0-7923-7253-0, pp187-196.

Sloane A (2002) "Methodologies for studying the home user" - World Computer Congress, Montreal, August 2002.

Sloane A (2003) "The Internet in the home: Changing the domestic landscape" in Stephanides, C. (ed.). Proceedings of HCII2003, Crete, June 22-27.

LONG-TERM VIABILITY OF SMART HOME SYSTEMS

Business Modelling and Conceptual Requirements on Technology

Greger Sandström, Stig Gustavsson, Stefan Lundberg, Ulf Keijer, Stefan Junestrand

Royal Institute of Technology, The Architectural School, Information Technology, Stockholm, Sweden

Abstract: In Stockholm during the years 1999-2002 some residential housing units provided with advanced IT-based functions were developed, also called smart homes systems. The purpose was to offer the residents an augmented living environment mainly regarding security and comfort. Only the IT company that originally developed the basic system fully knows it in depth, a fact which has shown to become very unfavourable for the residents. The running cost, also for minor changes of the functionality of the system, has turned out to be high and prohibiting. Also the contracted regular maintenance requires access to unique competence, which also has become costly. Recently the company has indicated a replacement of the smart homes system with a simpler one based on the Internet. Business models for viable home systems should include the occupancy phase, which to date have been neglected. A crucial issue is who will and can assume long-term responsibility for surveillance, maintenance, and added functionalities over time. This paper will discuss some principal questions and prerequisites regarding business viability in relationship to smart home systems, and its significance for business modelling.

Key words: Smart homes, sustainability, design, systems.

1. INTRODUCTION

In the past five years a number of smart home housing projects were developed in Sweden. During the subsequent post occupancy running phase questions have been raised regarding maintenance, repair and upgrading of the installed smart home functionalities. Maintenance and running costs are in general not important issues in the construction phase of a housing project, let alone energy conservation. Also in these projects the long-term issues were by and large neglected during the planning phase, including the smart home systems. This paper will discuss some issues regarding maintenance of smart home systems and related technological requirements, based upon the experiences from some smart homes developments, now in their occupancy phase.

Below a brief description is given of the smart homes concept and of principal roles and actors forming necessary constituents of the construction process. A case study will be outlined. Further, some typical information entities to be handled by an integrated smart home system are illustrated and it is discussed how this information could be preserved over time. To conclude the presentation possible business actors for the development and maintenance of smart home system are introduced.

1.1 Smart Homes

In traditional homes a considerable number of principally autonomous systems and networks support daily life, e.g. telephony, access control and security, household appliances, television, PC (data) networks, audio and video information and entertainment systems, lighting, electricity, ventilation, heating, air-conditioning, water supply and more. Generally, these systems are not connected or functionally integrated into each other. However, the connection of single household appliances and systems to networks and more complex systems has begun. Its technological and functional roots are found in *intelligent buildings* that arose and became important for the building industry in the 1970s. Intelligent buildings initially focused on commercial premises, i.e. offices and industrial buildings, providing technology for sensing and control, primarily out of a building management perspective. When the discussion turned from operation (of buildings) to living in buildings (homes), the term *smart homes* became generally used.

Smart homes are homes provided with some integrated technological system offering remote or central controlled functionalities and services[3]. In

[3] Example of services that can be created by the integration of the different systems are for example if a water leakage is detected at the laundry machine in the kitchen when nobody is at home, the water supply to the home is automatically turned off and telephone calls are

a smart home the inhabitants' desires and needs concerning all or a (substantial) part of these aforementioned equipment and functionalities are in focus. In principle, the technological infrastructure is based on the integration of the systems for home automation, security, multimedia and telecommunication. The systems are connected to each other and to the outer world through a residential gateway, see figure 1. Recently, it has become obvious that wireless connections may be just as viable as an alternative to wired solutions. The residential gateway could in this case be located outside the home.

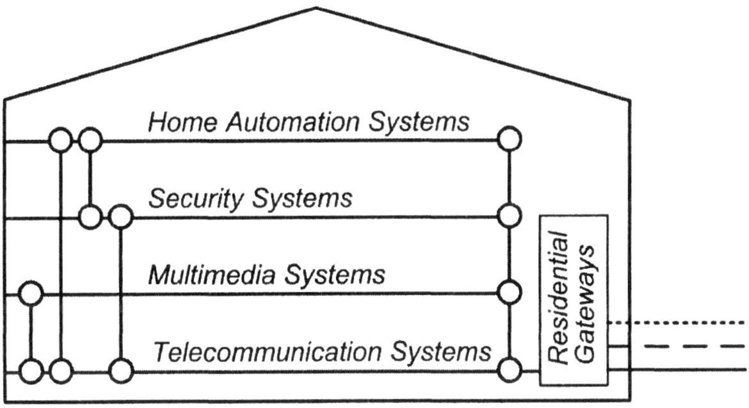

Figure 1. A schematic model of a technical structure of a smart home (Junestrand, 2004).

1.2 Models for organisation and business

An organisational model represents a company's total operation or a specified part of it. It usually takes the form of a graphic depiction of the structure and activities, accompanied by an explanatory legend or full text. The operation could likewise be project undertakings of various kinds, e.g. construction projects. There is a multitude of modelling tools available in order to describe organisations, their tasks and their business operation (see e.g. Barnley et al. 1986; Tarandi 1998 citing Mayer 1994). In this paper a task model related to the establishment of smart homes will be discussed and

generated to the pre-programmed telephone numbers, an SMS or an instant message is sent. Or when all the family has gone to bed a push-button upstairs will turn off the lights, bring down the blinds, lower the temperature 5°C and turn on the alarm. At 7 a.m. the next day the blinds automatically rise, the lights in the hallway and kitchen are turned on, the alarm is turned off and the temperature has been brought up to the normal 21°C.

how this model would support also the occupancy phase. Maintenance over time is a key factor for a viable development and marketing of smart homes on a broader scale.

As all business ends up with the customer, the first question asked is:

- "How is a viable organisational and business model designed in order to create prevailing added value for the residents?"

Further

- "What are the tasks and what are the actors' roles when included in the model? What are the requirements on the actors so to fulfil these roles?"

and

- "Are there essential features of a viable home network system to facilitate the tasks?"

This paper should be considered as an embryonic attempt to understand a fundamental problem of how to secure a sound development of the smart home business for the next few years lying ahead.

2. THE STUDY OBJECT

In the heart of Stockholm, a smart home residential housing unit, called Vallgossen was constructed. First occupancies took place in September 2001. Vallgossen comprises 126 flats in total. Their sizes vary from 44 to 144 square meters.

Figure 2. The Vallgossen project

In Vallgossen all 126 flats are equipped with a basic smart home system called "type 1 ", while 21 flats have an additional, more advanced system, "type 2 ". Two additional flats were test units for trying out assisted living, and equipped with some complex equipment for rehabilitation purposes (Sandström and Keijer, 2003).

The smart home system in each flat is accessed via a laptop computer. An important objective when designing the smart home system was to make its physical appearance as invisible as possible. Another objective was to make it user-friendly. It should not be necessary to control the system in order to live well in the flats. The smart home system provides the residents with information about security, energy consumption and available facilities, e.g. for booking the common laundry machines. It includes a family calendar and ordinary e-mail. It is possible to leave e-notes to other members of the family and to create a common reminder list.

The basic idea was that the smart home system should play a central role in the residents' everyday life. The residents would be able to gather important everyday family information in one place, e.g. to communicate with each other (e-notes), to communicate with their friends (e-mail and address book) and to co-ordinate their activities (calendar). Further, it would help them to feel safer at home (alarms) and to have control over their energy use. The availability of appropriate technology at the time defined the functionality of the installed systems. Noteworthy, when this building system was planned there was limited knowledge of the consumers' opinions and their perceived valuation of the available smart functions.

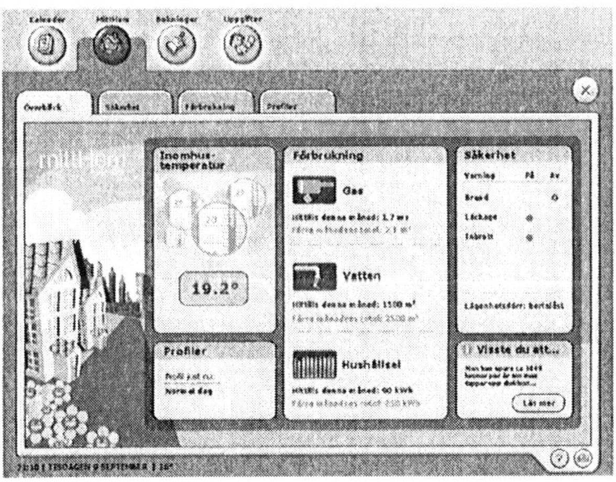

Figure 3. Example screen shot of the home network

The smart home system has been evaluated with regard to the residents' opinions (Sandström et al., 2003). It was found that the IT functions were not a conclusive reason for acquiring these flats; the households considered them rather as a bonus. Still the IT functions represent a habitation value. Functions increasing safety and security (e.g. alarms) and saving time (e.g. booking of common facilities) were the most appreciated functions. The possibility of energy conservation and of access to separate non-integrated functions was less appreciated. In a recent survey the residents in Vallgossen were asked "Specify to what degree you would miss the following functions if they are not available in your next home?". The results regarding some of the functions are found in table 1.

Table 1. Residents' opinions regarding smart homes functionalities

	very small degree / not at all	*small degree*	*high degree*	*very high degree*	*no opinion*
Fire alarm	-	1 %	5 %	94 %	-
Burglar alarm	-	2 %	14 %	84 %	-
The "away-lock"	6 %	1 %	22 %	71 %	-
Electronic keys	1 %	9 %	30 %	60 %	-
Bookings of common facilities	9 %	16 %	28 %	46 %	1 %
Individual energy measurement with graphical presentation of consumption	24 %	45 %	18 %	10 %	3 %

3. THE SWEDISH CONSTRUCTION MARKET

Many parties are involved in a typical housing construction undertaking. Many different roles are identified. In this article the client is defined as the developer of the undertaking. In particular this means that the client-developer takes the ultimate financial risk. The client is the body (person or company) that defines the building product, procures all site work and deliveries, monitors the construction work and is responsible for its proper completion. The client obtains the building permit and takes the prime responsibility for fulfilling any rule with respect to the undertaking. Especially, in this case, the client-developer purchased the entire smart home system from an IT system provider. The latter was backed by leading Swedish appliances and telecom industries.

Housing construction is carried out both by public and private companies. To a large extent the private construction companies transfer the completed building to some final owner or manager.

In Sweden there is a special type of residential housing, particularly in focus for this paper, where someone pays down a lump sum and acquires membership in a housing association, which has acquired the building. With the membership is associated a legal right to occupy a specific predefined flat.

The responsibility for the proper function of the building rests with the developer as a two-years warranty. This warranty is based on a long experience within the building industry. However, for complex structures, like smart home systems, such a limited warranty has shown to be insufficient. For the Vallgossen project the warranty period of the smart homes functions was extended another three years, see figure 3. As will be obvious below, the extension did not remedy the topical problem significantly. It postponed the pop-up of occurring problems a few years, nothing else.

It is clear that the end-users, the residents, have very little influence during the planning and construction process. Of course, this may cause problems already for traditional housing occupancy. These problems may increase considerably when a smart homes system is added.

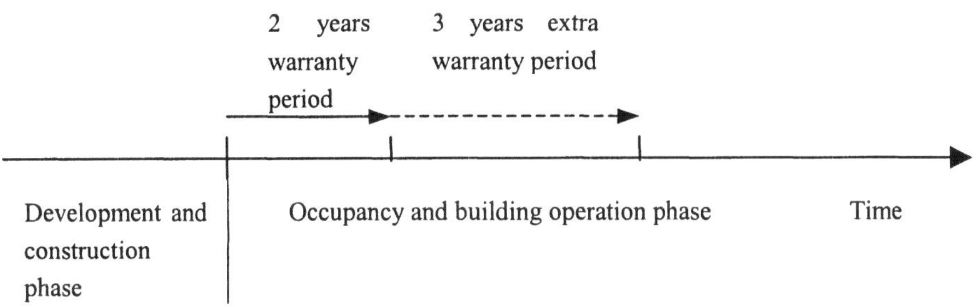

Figure 4. Extended warranty period, exclusively for the smart home

Already two years after the occupancy, the IT system developer closed down its activity and outsourced maintenance and support of the smart home network system to another company. Some employees, with unique competence, were transferred to this new company.

4. ORGANISATIONAL MODEL FOR THE VALLGOSSEN DEVELOPMENT

The developer initiated the idea of introducing smart homes in Vallgossen. The market for smart homes was growing and the client had to take some new initiatives. However, the available IT systems on the market were based on ideas and prototypes. The developer had not only to sign contracts with entrepreneurs (which he is used to) but also to define demands on the IT system, its design and functionality.

Table 2. Roles and dependencies in the construction process including smart home functions

Role	Code	Task	Dependencies	
			Dependent on	Controls/Co-ordinates
Client-developer	C	Organises the whole undertaking. Takes the financial risk until the homes are fully transferred the owners.	EU	All, except EU
Project Management	PM	Co-ordinate tasks between the client-developer and consultants, subcontractors and providers.	C	All, except C and EU
Architect	A	Principal consultant (in Sweden, however, with less superiority than in most other countries)	PM	CS, CM, CE, CSH
Consultant (structural)	CS	Develops blue prints for the structure in the building.	PM, A	
Consultant (mechanical)	CM	Develops blue prints for HVAC system in the building.	PM, A	SM
Consultant (electrical)	CE	Develops blue prints for the electrical system in the building.	PM, A	SE
Consultant (smart homes)	CSH	Co-ordinates IT installations between consultants, contractor, subcontractors and providers.	PM, A	SSH, SM, SE, PAP, PC, PAC
Subcontractor (mechanical)	SM	Install HVAC on the construction site.	PM, CM, CSH	
Subcontractor (electrical)	SE	Install electrical wires on the construction site.	PM, CE, CSH	
Subcontractor (smart homes system)	SSH	Install the IT system on the construction site.	PM, CSH	
Provider (appliances)	PAP	Delivers appliances to the construction site.	SSH	
Provider (windows, doors, etc.)	PC	Delivers windows, doors, etc to the construction site.	A, SSH	
Provider (access systems)	PAC	Delivers the access system to the construction site.	A, SSH	
The end user (the resident)	EU	The residents occupying the flats.		

The different roles necessary for the organisation of the planning and realisation of the homes are the client-developer, the architect, a number of professionals-consultants and subcontractors, see Table 1. The table could be extended, however, it is considered to be sufficient for the purpose of this article. Note the two roles, the *smart home consultant (CSH)* and *the smart home system provider (SSH)*, both representing entirely new tasks to be fulfilled in the construction process.

The smart home system subcontractor's commitment (see table 1) is primarily characterised by the long-term agreement with the client-developer, in Vallgossen five years, figure 2. The smart home system, which is a part of the service agreement, contains both software and hardware. Due to fast development of ICT, uncertainty concerning the future needs of maintenance and costs of the smart home system will be intrinsic. Contractual conditions putting demands on both the installed home system and on the agreement as such are difficult to formulate unambiguously. Especially, the end users' best interest is difficult to maintain over time. So called *agent theory* may offer additional perspectives useful for the treatment of uncertainties between the principal and the service provider (the agent), including contractual provisions and terms of reward (see Jensen & Meckling, 1976; Eisenhardt, 1989).

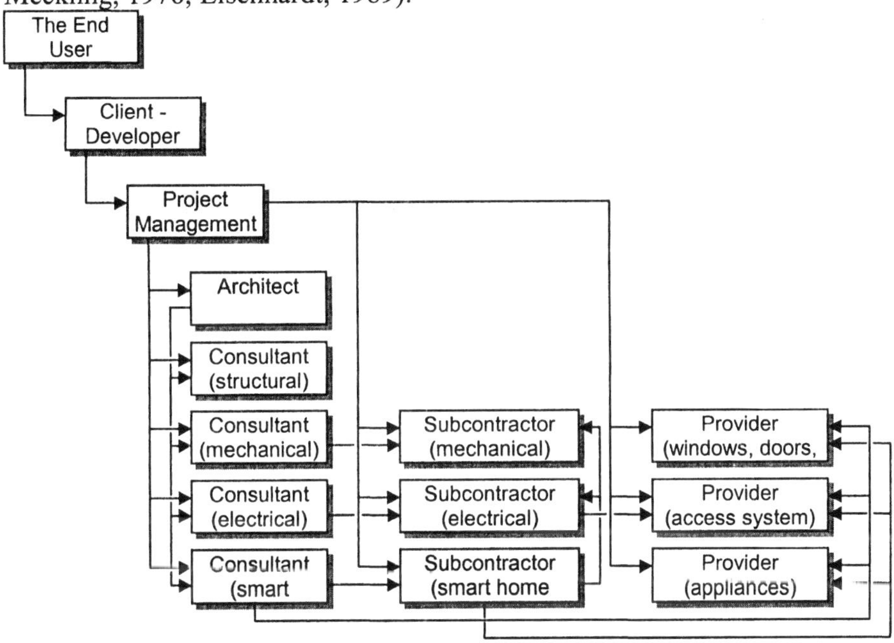

Figure 5. Co-ordination and dependencies in the building process including smart homes (IDEF0 notation).

Table 2 is largely mapped into a semantic schema depicted in figure 5. It is obvious that the introduction of the smart home system has increased the interdependencies between the different roles required for the proper construction of houses. Dependencies are inconvenient in a smoothly running construction process. Failures and remaining deficiencies are very often referred to lacking co-ordination between subsequent or parallel tasks of the process. The general problem is elaborated by for example Scherkenback (1991) when describing Deming's 14 quality assurance principles. For the construction process contributions by Sörensen (2003) and Appelqvist and Keijer (1994) could be referred to among many others.

5. THE OCCUPANCY PHASE

5.1 The away lock

When the entrance door is locked with the away-lock, the water supply is automatically shut off two hours later, as well as gas or electricity to stove and sockets. Further, the ventilation in the bathroom is reduced (if the moisture detector does not indicate need for ventilation) and the burglar alarm is activated. When the resident returns the lamp in the hall is lit and the main water supply, the gas, the sockets and the ventilation are turned on automatically. The lamp in the hall is on until the resident switches it off manually.

5.2 Typical events

Significant events have been noticed during the early maintenance phase at Vallgossen. Two examples connected to the so called "away-lock"-function is described below.

Example 1: The automatic hall lamp broke in one flat. The cost to repair the lamp ended up at some 880 Euro, including error detection in software and replacement of vital hardware (which were not manufactured anymore). The resident was not willing to pay for the repair and it is still dark in the hall.

Example 2: Magnetic valves control the main water supply in the flats. The valves are connected to the away-lock. The residents have now, during the maintenance phase, discovered that many valves are out of order. In some flats all valves are broken. In some other flats only a single valve does not work. The crucial question that has arisen is *when* the valves stopped functioning – during the maintenance phase or as a result of a software error from the very beginning. No checking procedure is available via the smart home system if the valves are in order or not.

5.3 Pulling out of responsibility

The home network was developed during the IT boom in the end of the 1990s. The smart home system provider (the IT-company) held large visions about the future market of their system. Their visions have not come true. Today the maintenance of the home network has turned out to be a heavy burden to the company. The company now wants to replace the home network with an Internet-based system, albeit the warranty period is not yet terminated.

Table 3. Representation of status vector for a refrigerator (assumed example).

Status vector	Description	Unit	Format
Temperature (upper space)	Temperature in refrigerator upper space	°C	format_t
Temperature (lower space)	Temperature in refrigerator lower space	°C	format_t
Interior lamp	Function/no function	logical	format_x
Interior lamp	Last exchange	date	format_d
Energy consumption record	Describes how information is represented	'info'	format_f
Energy consumption vector[4]	According to description	kWh	format_f.e
Door open[5]	Open/not open	logical	format_x
Door open record	Describes how information is represented	'info'	format_f
Door open vector[6]	According to description	'time'	format_f.1

Obviously, the smart home market presently does not offer any viable long-term return opportunity. A sound attitude to the operation phase, with a time horizon of a decade, is required. Elaborate business models can hardly be applied to remedy this precarious situation without rethinking the basic technological requirements when applying ICT for the whole business. Smart homes should be regarded as a service to end users. A general aim for consumer products is to be able to apply the product effortlessly, without prior instruction or training ("plug-and-play"). The same principle must be the ultimate goal if effective business models would be able to establish long-term living in smart homes. Some initiatives are referred to Newmarch (2000), Coulouris et. al. (2001) and OSGi (2004).

Let us turn to another example in order to elaborate the issue a bit further.

[4] Recording of energy consumption of the refrigerator could be summed up with the household's total energy consumption. It could serve as an alarm for a defect compressor performance.

[5] Could alert an alarm (on TV, on mobile phone, in home, in a relative's home)

[6] Could forward information on activity performed during a specific period (e.g. at supervision of old people)

6. EXCHANGE OF REFRIGERATORS SEEN AS A SMART HOME PROCEDURE

6.1 Status and control information

Any equipment or function in a smart home is defined by a vector showing its status (Table 3), currently or over time, and another vector (Table 4) defining the control spectrum of the item.

Table 4. Representation of control vector for a refrigerator (assumed example).

Control vector	Description	Unit	Format
Set temperature (upper space)	To specify temp. in refrig. upper space	^{o}C	format_t
Set temperature (lower space)	To specify temp. in refrig. lower space	^{o}C	format_t

6.2 Replacing the refrigerator

The old refrigerator is supposed to be integrated into the home network system offering the functions according to table 2 and 3. The new one has to have similar properties if the home network system will offer the resident the same benefit and if no unwanted effects are to occur. The crucial parts are the physical connections and the understanding and interpretation of data that are to be exchanged with the new unit.
If we disregard the physical connection (let's assume RS232), the data representations are supposed not to be identical between the two systems (refrigerators). In principle, any of two situations may prevail, either the appliance manufacturers adapt to existing home network systems or the home systems on the market can handle all occurring refrigerators.[7]

7. DISCUSSION

Large demands are put particularly on two specific roles in figure 5, i.e. the project management and the smart home consultant. In the Vallgossen undertaking the subcontractor of the smart home system showed to be no

[7] C.f. the way MS-Windows technically handles peripherals to be connected to PCs. Unfortunately, the situation on the home system market cannot be compared with the computer market.

guarantee for the long-term maintenance. So, the question remains: "Do we have any candidate who can find a viable business opportunity by taking this long-term responsibility of smart home systems?" Hardly, anyone is to be found in the traditional construction process. A look at different service providers may reveal some opening. Tentatively, three marked actors are identified, which may be in the position, having interest in the existence of viable smart homes systems and, however less certain, having the necessary competence for maintaining long-term activity in the field. For the time being most in focus are 1) security systems providers, 2) providers of comfort and energy, and 3) caring providers.

Security companies could take this role since they already provide access systems to buildings. A security system could form a platform which would be able to integrate other subsystems and appliances.

Providers of climate comfort and energy systems already serve many commercial buildings with different systems of measurements and controls, which would be possible to adapt to single family houses and to individual flats in multi-family buildings. These systems could very well be extended to other smart homes functions, and form a basis for truly integrated systems.

Some large companies provide home care associated with technical support systems, which probably will become more common in the future.

These three categories of companies all have an interest in offering added value to the home. Generally, they also have a functioning administration and – apart from the home care providers – a technical tradition. These companies may have an interest in providing more integrated home systems, and they could fulfil the role as smart homes service provider.

7.1 Requirements on the smart homes service provider

The service provider needs a technical infrastructure with IT-systems that automatically connects all smart home systems at a distance. Necessary staff is required that can manage electrical installations, electronics and informatics for basic configurations (reconfigurations) on the site.

Activities during the running phase that the service provider should be able to perform are:

- Remotely check the status of the system.
- Become informed automatically of any dysfunction or defect of the system.
- Maintain and upgrade the system on site on a regular basis.
- Be able to assist on the site with minimal delay.
- Give around-the-clock support on-line and by telephone.

Still, the schema lacks one actor: the *local technical service provider*. The service provider monitors households in a large geographical area and cannot have service technicians available 24 hours a day seven day a week. Here the local technical service provider has a role to fulfil. If something in a home system fails, the resident should obtain support immediately. But the local technical service provider will be placed in a difficult position. He will probably not be able to influence the home system in its design and construction phase. He will be contracted by the end user and must deal with the home system regardless of its design and functions. One possibility is that the subcontractor of the smart home system has service agreements with a number of local technical service providers. In this situation the local technical service provider will have access to expertise knowledge about the smart home system. Another possibility is that all subcontractors' of smart home systems regularly educate local technical service providers about their specific home system. Figure 5 indicates the local technical service provider position between demanding customers/residents and a complex technical system (the smart homes system).

Another important issue to be aware of is the interconnection between the smart home system and Internet. If the end user contracts a service provider for maintenance of the home system, it will probably be monitored from a distance. This situation will put special demands on the home systems security. It is desirable to separate the smart homes system from the Internet. Otherwise, firewalls and virus programs must be included in the system. This situation will put a large responsibility on the resident, e.g. virus programs must be regularly updated. And still, it cannot be guaranteed as a 100% secure system. A smart home connected to Internet will always be exposed to external threats like hackers and viruses. On the other hand, if the home is separated from the Internet large demands will be put on equipment providers and smart homes systems to agree on far-reaching compatibility of components and parameters.

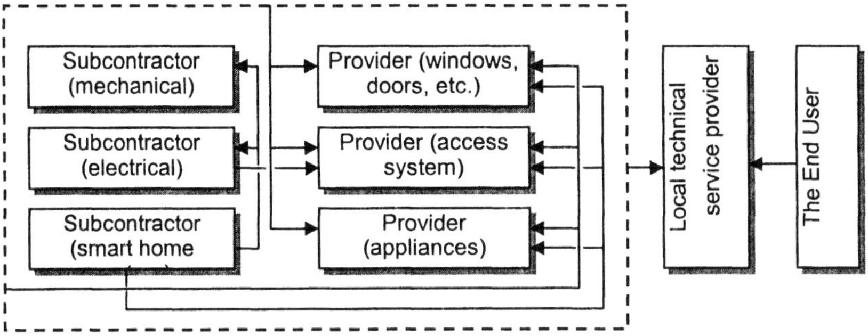

Figure 6. The local technical service providers place in a semantic schema.

ACKNOWLEDGEMENT

This work is funded by the Swedish Research Council Formas and the Development Fund of the Swedish Construction Industry (SBUF) which is duly acknowledged.

REFERENCES

Appelqvist I. and Keijer U. (1994). *Building Integrity. Interaction between Building Parts, Systems and the Actors of the Building Process.* CIB W78 Workshop on Computer Integrated Construction, August 22-24, 1994, Espoo, Finland.

Barnley J. B. and Ouchi W.G. (eds.) (1988). *Organisational Economic*, 2nd ed. Jossey Bass Publishers, San Franscisco, Cal.

Coulouris, G., Dollimore, J. & Kindberg, T. (2001). *Distributed Systems – Concepts and Design*, 3rd Ed. Addison-Wesley, UK.

Eisenhardt, Kathleen, M. (1989) *Agency Theory: An assessment and review.* Academy of Management Review. Vol. 14:57-74

Jensen, M. C. and Meckling, W. H., (1976). "Theory of the Firm: Managerial Behavior , Agency Costs and Ownership Structure." Journal of Financial Economics, 3: 305-360

Junestrand, S. (2004). *Being private and public at home - an architectural perspective on smart homes.* TRITA-ARK Avh. 2004:3. School of Architecture, Royal School of Technology, Stockholm, Sweden.

Mayer, R.J. (ed) (1994). IDEF0 Function Modelling. A reconstruction of the original Air Force Wright Aeronautical Laboratory Technical Report AFWAL-TR-81-4023, Knowledge Based Systems, US.

Newmarch, J. (2000). *A Programmer's Guide to JINI Technology.* Apress, California.

OSGi Alliance. (2004). *About the OSGi Service Platform.* http://www.osgi.org/

Sandström, G. and Keijer, U. (2003). *Alloggi domotici integrati e di addestramento per persone con problemi mentali* (Integrated smart living – Training flats for persons with acquired brain dysfunction). Abitare e Anziani Informa (Housing and Elderly Info), No. 1-2, pp 85-90.

Sandström, G., Werner, IB. and Keijer, U. (2003). *Smart Homes Evaluated.* Open House International, Vol. 28, No. 4.

Scherkenback, W. W. (1991). *The Deming Route to Quality and Productivity.* Road Maps and Roadblocks.

Sörensen, N. (2003). *Co-ordination of the Design and Building Process for Optimal Building Performance.* In Construction Process Improvement. Editor Atkin, B., Borgbrant, J. & Josephson, P-E. Blackwell Publishing, UK.

Tarandi, V. (1998). Neutral Intelligent CAD Communication – Information exchange in construction based upon a minimal schema. Construction Management and Economics, Royal School of Technology, Stockholm, Sweden.

A BLUETOOTH HOME DESIGN @ NZ

Four Smartness

Olaf Diegel, Grettle Lomiwes, Chris Messom, Tom Moir, Hokyoung Ryu , Federico Thomsen, Vaitheki Yoganathan, Liu Zhenqing
Massey University, Auckland, New Zealand

Abstract: Much of the work in the smart house technology has been done on individual technologies, but little has been done on their integration into a cohesive whole. The Bluetooth house project at Massey University in New Zealand, which was initiated in 2002, embraced a systems engineering approach to design a usable smart house, aiming at a complete and integrated solution, which can be customised, based on individual needs, to give elderly people independence, quality of life, and the safety they require. This paper presents how the Massey Bluetooth smart house design project has been carried out and what the smart home may look like in the near future. Considering current technical feasibility and the advances in other research, it is suggested that for a house to be considered as truly 'smart', four levels of smartness are imperative: *smart sensors*, *smart management*, *smart control*, and *smart appliances*. The Bluetooth house at Massey University incorporates these four smart technologies and allows all these individual technologies to be integrated into a seamless whole. For smart sensing, the project employed Bluetooth technology to connect the whole house, and to locate the user's position. In order to coordinate all the technologies, a smart management system was developed, that is capable of coordinating the information for commands, feedback from smart appliances, and user's location information. It can make intelligent decisions on what to do, or relay necessary information to individual intelligent devices throughout the house. In addition, the medium of communication with the house must be as natural as possible, in order to make it as easy as possible for the occupants of the smart house to interact with and the various smart appliances. A voice-activated universal remote control and a new microphone system are being developed to this end. Finally, the smart house has to provide an enjoyable experience that can promote the uptake of smart house technology by users in the future. An interactive TV environment is being developed to this end. The Massey Bluetooth house project is not so much aimed at a cutting-edge technology in smart house design, but at integrating technologies into a seamless, cohesive whole through the application of four levels of smartness.

1. INTRODUCTION

Since Mark Weiser of PARC has coined the phrase 'ubiquitous computing', there have been great advances in this research. Much of the enabling technologies in this area, as predicted in his visionary paper (Weiser, 1991), are now to some extent available. For instance, Global positioning systems, Personal Digital Assistants, Bluetooth networks and Radio Frequency Identification networks are all applicable in realising the concept of a ubiquitous computing environment.

It is noted that these technologies have been great successes in each of their individual consumer worlds. A smart home (or smart house), however, asks them to be integrated to a level where the technologies and appliances in the house help make life easier, safer and more enjoyable for the occupants (Rogers and Mynatt, 2003). It poses the three important issues in that a systems engineering approach is needed to make all areas of the smart house work together seamlessly (Jacko and Sears, 2003); the smart house should be transparent to the people in the home (extended from Norman, 2001); and finally multidisciplinary cooperation is required to achieve these goals. It is also note that much work has been done on individual technologies that can be of help in caring for the frail and elderly, but little has been done on their integration into a cohesive whole. The Bluetooth house project at Massey University, which was initiated in 2002, follows this systems engineering approach to develop usable smart house technologies in New Zealand, with collaboration between engineers from electronics, robotics, telecommunication technologies, and psychologists.

Apart from the academic interest, the political and social aspects of New Zealand are also considered in this project. In 2002, the authorities of New Zealand Health sector initiated their strategic approach to provide appropriate health care to the elderly population in New Zealand (Health Sector Strategic Report 2002). The report concluded that New Zealand has a high and increasing elderly population ratio without any support from the other family members and suggested that rest homes with monitoring facilities would be very effective in taking care of this population segment.

The goal of the Massey Bluetooth home project is thus to create a complete and integrated solution, which can be customised, based on individual needs, to give elderly people independence, quality of life, and the safety they require.

2. CHALLENGES IN MASSEY BLUETOOTH HOME DESIGN

The Massey Bluetooth home project, presented here, offers a potentially high opportunity to demonstrate a basic level of smart house technology, focusing on two challenging issues: Integrity and Usability.

2.1 Integration

It can be seen that the key success factor of smart house technology is how well the individual technologies can be integrated to provide a comfortable life in the home. Integration is thus what smart house researchers are ultimately aiming at. The current technologies available still need to be reconfigured for this objective, as there are many interdependency issues that arise as the individual technologies are integrated into a cohesive working application. For example, the Aware Home project at Georgia Tech (see more details in http://www.cc.gatech.edu/fce/ahri) and the project Aura at CMU (see more details in http://www-2cs.cmu.edu/%7Eaura) also focus on the integration of the individual technologies.

The Massey smart house team's approach is thus to identify the opportunities and limitations of current technologies in the home and to introduce a plausible solution for their integration through the use of a Bluetooth network. Some technologies are also being developed to ensure the integrity of low cost smart house technology in order to meet the market's demands.

2.2 Usability

In the context of work, the key components of usability are recognized as task fit and ease of learning. Current smart houses are often designed from a mechanical view so that poor ease of use and task fit are major barriers to the uptake of such ubiquitous computing technology.

Whilst conventional concepts of usability are equally important, they miss something about the nature of smart home environments, specifically, activities in the home. Smart house researchers presume that many activities in home do not have a clear aim or task objective and may be done simply for the enjoyment they provide. Thus, in the smart house the criteria for usability have mainly to do with the user's experience rather than the user's ability to complete some task (from personal communication with Monk, 2003).

The approach of the Massey Bluetooth home project is thus to identify design principles for ease-of-use and ease-of-learning, previously developed and applied in the workplace, to the problem of configuring and re-configuring networks of devices in the smart house. At the same time new concepts of usability will be identified, building on the work on this topic currently going on in the other ubiquitous computing projects at Massey.

The following sections discuss how these aims are being investigated in this project.

3. THE MASSEY BLUETOOTH HOME DESIGN

Considering current technical feasibility and the advances in other research, e.g., MIT Oxygen project, Home Automated Living and DELTA project, it is presumed that for a house to be considered as truly 'smart', it requires four things:

- Smart sensors: it needs to know who is in the house, where in the house they are, and what special needs or preferences they may have.
- Smart management: it needs a central management system, which is based on the occupants identities and locations, can coordinate all the smart appliances and devices in the house to best fit those occupants' needs.
- Smart control: a speech recognition system that allows the users to communicate with the house in a natural manner, without having to wear headsets or consciously have to activate a microphone, etc.
- Smart appliances: to be a smart home environment, it needs some smart appliances that have enhanced capabilities of the conventional home appliances. For instance, Internet-connected appliances are now within the financial reach of the ordinary consumer, resulting in a range of new services to enhance our lives. At the application level, what sort of smart appliances would be useful in the smart home environment is also an important concern.

The Massey Smart House incorporates these four core technologies to allow all these individual technologies to be integrated into a seamless whole. The following outlines the details of each sub-project.

3.1 Bluetooth Network with Bluetooth Watch

There are many workable networks in the smart house design, e.g., Wi-Fi, GPS, RFID and Bluetooth. One requirement in connecting the house is that the network has the ability to detect where users are within the network. That is to say, the house needs a dynamic network throughout the house that allows devices to communicate with the house management system that coordinates the information within the house, and the occupants of the house, based on the location of the occupants. This can be achieved through a variety of means including the technologies mentioned above. The indoor tracking system developed by AT&T Laboratories in Cambridge, for example, uses a network of ultrasonic modules to keep track of the users (Harter et al., 1999). The disadvantage with such a system is that the house will only react to occupants wearing the appropriate ultrasonic transmitters.

Bluetooth has the advantage of being an almost ubiquitous technology used in many common appliances such as cell-phones, PDAs, and more, giving it the advantage of having an already available range of transmitters, allowing the smart house to react to a wider range of people. Bluetooth has a range of around 10 meters, which is adequate for certain forms of communication. In contrast, RFID is capable of covering only a relatively short range (around 1 meter,) which is entirely dependent on the radio frequency and the power, but its speed of communication is faster than that of Bluetooth. The Massey smart house adopted Bluetooth as the communication technology in our project, because it allows devices to automatically talk to each other when they come within a certain range, at a relatively reasonable cost, and it is relatively easily extendable to allow communication throughout the entire house, e.g., up to 1000sqm.

Figure 1. Massey Bluetooth watch

In parallel with the Bluetooth network, a small Bluetooth device embedded in a watch satisfies the requirement of locating the user's position within the network. The project includes the development of a Bluetooth enabled watch/blood pressure monitor as depicted in Figure 1, as well as the design and construction of the Bluetooth ubiquitous network, and the network software.

The Bluetooth network consists of the Bluetooth watch and several small reduced-range Bluetooth transceiver modules that are attached to the ceiling of the room or house to form a grid of linked modules (see Figure 2). The modules are spaced 2 meter apart and each module is set to have a range of 2.4 meters. The grid is connected to a computer running software to deal with the information received from the network. This software includes the ability to track the user over a map of the house, and display the personal data contained in the users watch. As smart appliances, e.g., interactive TV, then begin to be developed and integrated into the system, this information can then be used to intelligently control the appliances.

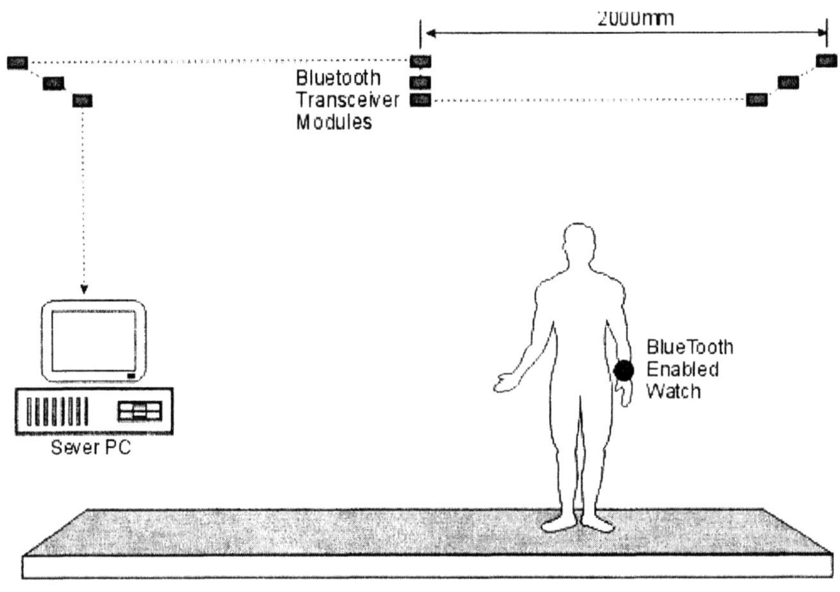

Figure 2. The Massey Bluetooth network

The Bluetooth-enabled watch is loaded with the users' personal data, and it has a range of 2.4 meters. As the watch is generally between 500mm to 2000mm from the floor (depending on the user's arm position), it is always within range of at least one receiver module but never in range of more than 4 receiver modules. It then becomes simple geometry to determine the location of the user relative to either one, two, three or four modules. This is well within the capabilities of Bluetooth, which is capable of communicating simultaneously to up to seven devices. An additional advantage of using Bluetooth is that standard Bluetooth enabled devices, such as cell-phones, can be used in tracking visitors in the house.

3.2 House Management System

In order to function efficiently and to avoid the duplication of much of the technologies being developed, the house requires a central computer and software package capable of using the information from the speech recognition system for commands, feedback from smart appliances, and user location information. Based on this information, the house management system makes intelligent decisions on what to do, or relaying necessary information to individual intelligent devices throughout the house.

The house management system involves PC based software that, based on the information received from the Bluetooth network, user commands via speech-recognition, and other sensors throughout the house, is capable of making intelligent decisions. Currently, the system is composed of two modules: an expert system to process the received commands and a conversation module that would operate as a 'Chatbot' to converse with the occupants in the home. Further modules such as a visual element to receive some forms of visual communication from the occupants are also envisaged.

Table 1. Massey smart house database.

Table name	Information held	Description
Commands	Command ID Command Degree of certainty	The commands table contains all the possible commands that the house management system can implement within the house. Each command has a unique command ID and a degree of certainty which indicates that the command received may be unsafe or needs to be confirmed.
Devices	Device ID Device	The device tables contains all the devices that may be accessed through the commands given by the user. Each device also has a unique device ID
Room	Room ID Room	The room table represent all the rooms (or physically isolated location) in the house such as kitchen or the garage. Each location also has a unique room ID
Commands-device	Command ID Device ID	The command device table links the commands and the device, i.e., 'light on' command in the universal remote control is linked to the device

Table name	Information held	Description
		'light'
Room-Device	Room ID Device ID	The room device table holds information about the devices that a particular room has within it.

For the working prototype, the house management system is made up of the house database which contains the commands and the rules to operate the devices. The current house database is represented in Table 1. As any process occurs it is checked against the database and the house management system passes the commands to Switching system so as to access appropriate devices.

3.3 Interaction via Speech Recognition

In order to make it as easy as possible for the occupants of the smart house to interact with and the various smart appliances, the medium of communication must be as natural as possible. Other forms of interaction styles, e.g., eye-tracking, may also be applicable for specific appliances or functions, but a good speech recognition system would resolve most of the traditional communication difficulties.

The Massey smart house therefore adopts a speech recognition system, which is capable of analysing spoken language and extracting necessary instructions from it. It involves a speech recognition system that allows the users to communicate with the house management system and a universal remote control for smart appliances, without the need for a headset microphone. The microphone system makes use of beam follower technology (see for details Griffiths and Jim, 1982), and a commercial speech-recognition software is employed, i.e., Dragon™ Naturally Speaking, for a universal remote control that is activated by voice.

3.3.1 Universal Remote Control

Too many remote controls are very problematic in a home, as the user generally intends to control only a particular appliance. That is where a universal remote control comes in. This allows the user to control all their appliances without using separate remote controls.

Yet, from a 'smart' perspective, the current universal remote control still needs to be advanced from its current interaction style based on keypad-input, to a more intuitive style using speech-recognition technology. A universal remote control is being designed, on which all commands can be activated with human speech rather than through the current user's keypad input on the physical remote.

Based on this account, a voice-activating universal remote control is being developed. The control flow in our design of the universal remote control is described as follows: First, Dragon™ Naturally Speaking recognises what the user said. Then, a program written in LABVIEW™ generates the related binary string which is the same as what the normal remote controller creates when it is clicked. Finally through the RS232 port, the binary strings are transmitted to a microcontroller and the corresponding pulses modulated at 38 kHz are propagated through the infrared LED. The schematic of this circuit design is shown in Figure 3.

In order to dim a light, TWS-434, RWS-434 and HT-12E, HT-12D are used as RF transmitter, receiver encoder and decoder, respectively. LS7631 is a dimmer chip which can adjust the fire angle of a Triac. A signal on the Din pin of HT-12D activates the oscillator which is in turn continuously checked by a decoder. The transmission is not valid until the received address is matched with the decoder's local address. The 555 circuit acts as a pulse generator which can generate different pulse widths which are fed to the LS7631. This voice-activated universal remote control is currently being tested in a laboratory environment.

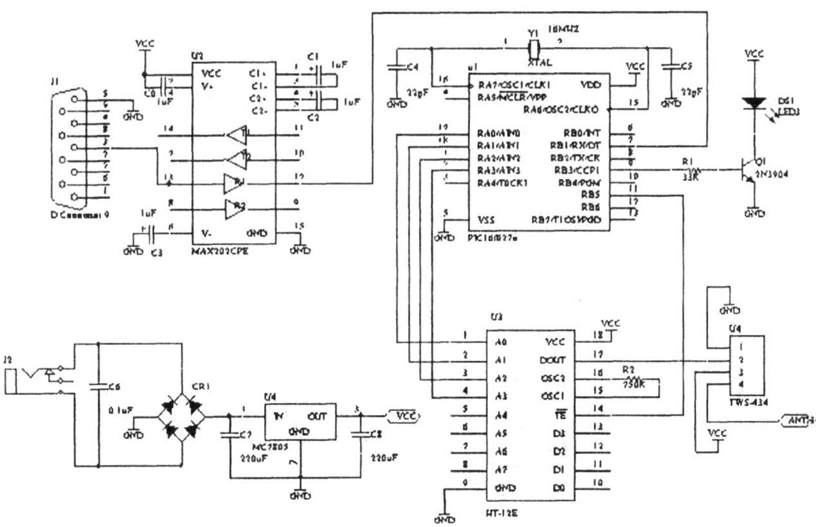

Figure 3. Circuit design of the universal remote control.

3.3.2 Microphone Array Beamformer

The same interaction style used with the universal remote control is also used to analyse the occupants spoken language and extract necessary commands for the house management system. A simplistic solution to this problem is to use a headset or body worn microphone to acquire the speech from the user. Yet, as the smart house technology is being aimed toward the frail and elderly population, this is not so practical. Furthermore, the majority of the occupants would prefer not to wear a microphone in their house at all times. Therefore, a microphone system is being developed that allows the users to move around freely and interact easily with the house management system by speaking naturally.

Using a directional microphone is a simple solution to this problem, but they seem to fall short of their objectives in adverse environments. The received signal from the microphone might include the desired speech signal as well as the background noise, which can consist of several components propagating from different sources such as computer fans, radio, TV, and other talkers. For the speech recognition system to work efficiently a signal to noise ratio of greater than 20dB is typically required.

Another potential solution to this problem is the use a microphone array, which will give the smart house users the advantage of really being able to move freely around the house. The microphone array makes use of beamforming technology to fight against the effects of the acoustic environment.

The selected adaptive algorithm to be used in the smart house is based on a modified version of Griffiths-Jim beamformer (Griffiths and Jim, 1982), which was originated by Van Compernolle and Leuven (Van Compernolle and Leuven, 1990). This algorithm has been demonstrated to perform well under noisy and reverberant conditions. The algorithm makes use of two adaptive filters based on Least Mean Square (LMS) (Widrow and Hoff, 1960). Since the LMS algorithm has some drawbacks with stability and selection of the step-size the system will instead be using an adaptive filter based on Normalised LMS (NLMS) (Haykin, 2002).

This speech beamformer makes use of two NLMS algorithms. The first NLMS is updated during a speech segment and the second NLMS is updated during the noise segment. The first one acts as an adaptive beam-steering filter and the second one acts as a filter for the noise. Only one of these NLMS algorithms is updated at a given time. The technique also uses a simple voice activity detector to analyse the received speech signal and determine if it is speech or noise. The corresponding NLMS algorithm is updated depending on the result obtained from this voice activity detector. The above algorithm is in the process of being implemented in real-time on a Texas Instruments digital signal processor.

3.4 A Smart Appliance: Interactive TV and User Experience

If, as discussed above, many of the activities in the home are undertaken for the enjoyment they can provide, i.e., watching TV and cooking, then the smart house has to provide an enjoyable experience that can promote the uptake of smart house technology by users in the future.

For instance, new television technology such as digital television might produce more pleasant experiences for housebound, disabled or elderly people, as it would allow them to access richer and more customisable information from their home. In particular, it is noted that the elderly are the biggest current consumers of television in New Zealand, watching on average more than 5 hours a day (National Statistics of New Zealand). This implies that they have less communicative involvement with their neighbourhood, and that they are at risk of becoming socially isolated from their community.

It is believed that our smart house, together with new television technology, can help lessen this social problem, as the technology facilitates social interaction in the community. In a similar context, Hampton (2003) has set up a wired community to see how much information and communication technologies facilitates community participation and collective action. Yet, the previous research does not propose the development of applications in the interactive TV environment, thus encouraging elderly people to adopt it. Following this work, the research team aims to investigate what kinds of applications in the interactive TV environment would facilitate interaction with their neighbourhood for elderly people, thus extending the concept of smart home into the smart community.

Based on this understanding, an interactive Java™ TV environment with the voice-mediated technology is being developed. The main functionality of it is to enhance the broadcast and viewing experience by providing such features as programming information and chatting with friends while watching TV programmes. This chatting facility can increase social ties between people, even while they are watching television. A prototype is being designed and implemented.

4. CONCLUSIONS AND FUTURE WORK

This paper presented how the Massey Bluetooth smart house design project has been carried out and what a smart home integrated with these technologies may look like in the near future. Currently, the Massey

Bluetooth house is only working in a laboratory environment. The house management system, the universal remote control, the beamformer microphone and the interactive TV are still being implemented or tested in this same laboratory environment. The evaluation of both the individual systems and the integrated system has been planned.

This paper has not discussed other smart house issues such as accessibility, emotionality, privacy, security, and sociality, along with the technological approach. We did not intend to trivialise these issues; however, as the current project aims at building a physical house for people to use as a model house for future living, the issues were not included in the immediate project, but will be studied as part of future work.

In conclusion, the Massey Smart House is not so much aiming at a cutting-edge technology in smart house design, but at integrating technologies into a seamless, cohesive whole, and drawing a picture of the technological home in the New Zealand environment as well as developing some business ideas with industry partners such as construction companies, appliance companies, and the government. This research is also intended to communicate the concept of the smart house to the public, encouraging people to access our facility, and thus feel the added benefits of integrating smartness into the home. In the end, the main beneficiaries of this project will be our elderly population who want to retain their independence, and their families and friends who can be secure in the knowledge that they are safe, well and comfortable. The health sector will thus benefit by being able to more effectively help and monitor people in their care. There will also be flow-on benefits for the construction industry, appliance industry, and for people who wish to improve their quality of life.

REFERENCES

Griffiths, L. J. and Jim, C. W. (1982) In *IEEE Trans. Antennas Propagation*, pp. 27-34.

Hampton, K. N. (2003) *Grieving for a lost network: collective action in a wired suburb, The Information Society,* **19,** 417-428.

Harter, A., Hopper, A., Steggles, P., Ward, A. and Webster, P. (1999) *The anatomy of a context-aware application., ACM/IEEE International Conference on Mobile Computing and Networking*, 59-68.

Haykin, S. (2002) In *Adaptive Filter Theory*, (Ed, Kailath, T.) Prentice Hall, Inc., Upper Saddle River, New Jersey.

Jacko, J. A. and Sears, A. (Eds.) (2003) "*The Human-Computer Interaction Handbook: Fundamentals, Evolving Technologies and Emerging Applications*", Lawrence Erlbaum Assoicates Publishers, Mahwah, NJ.

Norman, D. (2001) In *Communications of the ACM*, pp. 36-37.

Rogers, W. A. and Mynatt, E. D. (2003) In *"How can technology contribute to the quality of life of older adults?"* (Ed, Mitchell, M. E.), pp. 22-30.
Van Compernolle, D. and Leuven, K. U. (1990) In *IEEE International Conference on Acoustics, Speech, and Signal Processing, 1990. ICASSP-90.* Albuquerque, pp. 833-836 vol.2.
Weiser, M. (1991) *The computer of the 21st century, Scientific American,* **265,** 66-75.
Widrow, B. and Hoff, M. E. (1960) In *IRE WESCON Convention Rerecord,* New York, pp. 96-104.

SMART HOME CONTROL VIA PDA

An Example of Multi-Device User Interface Design

Brigitte Ringbauer
Fraunhofer Institute for Industrial Engineering, Stuttgart, Germany.

Abstract: Smart home systems must be controlled comfortably and securely by a broad range of users and in diverse contexts of use. One possibility to meet the requirements deriving from these diversities is to provide multiple devices to control the system. When a smart home application is used by one person on different devices, the user interface concepts should support an easy transition between devices. This article describes the development and result of a multi-device user interface to control a smart home system with focus on usability issues. The target devices are touch screen display and PDA. The design principles for inter-usability proposed by Denis & Karsenty 2004 are discussed and adapted with respect to a balance of inter-usability and usability of the specific device interface. The refined PDA user interface is described in detail.

Key words: smart homes, PDA, control applications, HCI

1. INTRODUCTION

Developing smart home solutions, we have to face a remarkable diversity in user and context characteristics. Within a household typically different social roles exist and accordingly household resources are used in a different way. The user characteristics may differ in gender, age, and experience - the latter two also change over time. The context of use of household appliances also differs quite a lot. A user interface has to support hectic and unfocussed behavior as well as effective task completion. The challenge is to fulfill these requirements from the user interface side and to provide a usable interface to system control and interaction. An example of multi-device user interface design (Seffah & Javahery, 2004) with focus on usability issues developed in the research project LIVEfutura[8] is described here.

2. USABILITY ISSUES OF MULTI-DEVICE SYSTEMS

The usability of a multi-device system comprises the usability of a concrete implementation of an application on the target device and the usability of the multi-device system as such.

There are many styleguides that provide principles for UI design for different platforms and devices (Lynch & Horton, 2001; Microsoft Corporation, 2001; Sun Microsystems, 2001; palmsource™, 2004; Windows 2004). Therefore this paper concentrates more on inter-usability issues (Denis & Karsenty 2004), i.e. usability issues that arise when an application is used via different types of devices.

Denis & Karsenty (2004) provide a conceptual framework for inter-usability of multi-device systems. They propose two dimensions that have to be fulfilled by a multi-device system to be usable given inter-device transitions: *knowledge continuity* and *task continuity*. Knowledge continuity means that what the user has learnt from the usage of an application via one device also holds for another device in the multi-device system. Task continuity has to do with a shared memory of user actions with the system that allows a user to continue tasks begun via one device via another.

To ensure task continuity when a device transition is made, all devices have to have the same information status and have therefore to be "always on". This is also a requirement coming from the fact that a smart home system is normally a multi-user-system. The first precondition to avoid problems with conflicting actions of different users and conflicting system- and user-actions is to provide real-time information on the status of the house on all devices (Ringbauer & Hofvenschiöld, 2004). In our study described here we did not develop a solution on the technical side for our user interfaces[9]. Therefore this article concentrates on knowledge continuity.

Denis & Karsenty (2004) suggest three main design principles constituting the basis of inter-usability:

[8] LIVEfutura (www.livefutura.de) was funded by the German Ministry for Education and Research, funding no. 01AK931.

[9] The touch screen display user interface was prototypically implemented as one user interface of an integrated smart home demonstrator from Fraunhofer FOKUS and Fraunhofer IMS in the project LIVEfutura.

- Inter-device consistency (perceptual, lexical, syntactical, and semantic consistency)
- Transparency what the system can do and how it works
- Adaptability, e.g. to the user profile

The principle *inter-device consistency* takes up aspects of the dialogue principles four and seven of the ISO 9241-10 (ISO, 1996) "conformity with user expectations" and "suitability for learning". The user expectations in the case of a multi-device system in the context of smart living environments are built upon:

- experience with non-technical applications and services in real life (e.g. pin board as communication centre in the home, experience with temperature setting on the radiator)
- expectations on the use of technology derived from prior experiences with information technology (e.g. office software, mobile phone, remote control)
- expectations on the interaction with one service provided by the smart home application from the usage of another service (e.g. remotely controlling the coffee machine and retrieve the fuel level of the car)
- expectations on the functioning of an application because of experience with other applications via the same device (e.g. a regular PDA user would expect an application for integrated home control to work similar to his organizer application on the PDA)
- prior experience with other devices of the same system (inter-usability)

The claim for transparency has to be fulfilled with a sound user interface concept on the respective device. In our study we focused on the issue of *inter-device consistency* and its relationship to "suitability for the task" (ISO 1996). Especially the points *semantical consisteny* and *syntactical consistency* may contradict the aim of efficient task completion on a particular device sometimes.

3. DEVELOPMENT OF THE USER INTERFACE CONCEPT

Most of the results presented here were achieved in the research project LIVEfutura. It was a project dealing with the specification and prototypical implementation of an integrated overall home control concept covering all relevant application areas: home automation, audio/video control, home appliances, home computer network, home telecommunication network, and even the private car. LIVEfutura covered both, the interconnection of the different subnetworks via gateways and proxies and the realization of integrating user interfaces for overall home control. The goal was to provide an easy, secure and satisfying access to smart home services for all.

For the development of the smart home control a user-centred design process (ISO, 1999) was adopted. The methods and results are described in more detail in Ringbauer et al., 2003.

3.1 Development of the information structure, and the basic interaction steps

The first step in the development process was to collect the functional and non-functional requirements through interviews with potential users. We were also living in the INHAUS Duisburg[10] for a two day period. The INHAUS is a fully connected home run by our project partners and the industry and is probably the most known smart home in Germany. Diary methods were used to capture the use of technology in the real living context (Ringbauer & Hofvenschiöld, 2004). In the requirements analysis it turned out that the user group is very heterogeneous and that we have to cope with a broad range of usage needs (e.g. activate the setting "leaving home" and remote access to security settings) and contexts of use (e.g. cosy Sunday morning vs. quick breakfast on a weekday). Therefore we decided to develop two types of interaction devices for the integrated smart home environment: a stationary device as a main control device and a mobile device that might be personalized.

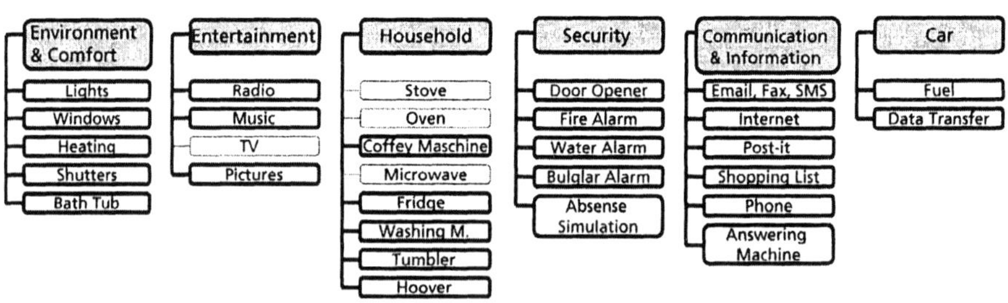

Figure 1. Functionally oriented information architecture on the basis of the card sorting (the appliances in the gray boxes have not been networked at the time of development)

For the stationary input device we chose a rectangular touch screen display. A touch screen display has the advantage to enable direct interaction with the finger being the pointing device. Having a pointing device makes it easier for inexperienced users. A touch screen display ideally placed in the

[10] www.inhaus-duisburg.de (English version available)

kitchen or the hall has been widely accepted in several studies (Kalmer, 2001; Meyer, 2002). Especially to fulfil the need of exchange of household information the kitchen seems to be a good place for a stationary device (Harper & Shatwell 2003).

In the first concept step we described the interface at a very general level. With the information collected and structured by scenarios (Carrol, 2000), extant systems analysis, task analysis (Kirwan & Ainsworth 1992), card sorting (Usability Net, 2003) and wording tests we built the basic information architecture, defined main screens and which steps are necessary to fulfill a task. Then we added the wording to application categories and services. The first information architecture is shown in fig. 1.

As in the *Unified User Interface Development* presented by Stephanidis & Savidis (2003) this basic interface structure with elements and actions stays the same for all device user interface concepts. In contrast to Stephanidis & Savidis we did not specify all possible interface artefacts with adaptation rules to enable an (automated) adaptation to all possible user and context variables. We chose two promising devices, touch screen display and PDA and adapted the user interface to them.

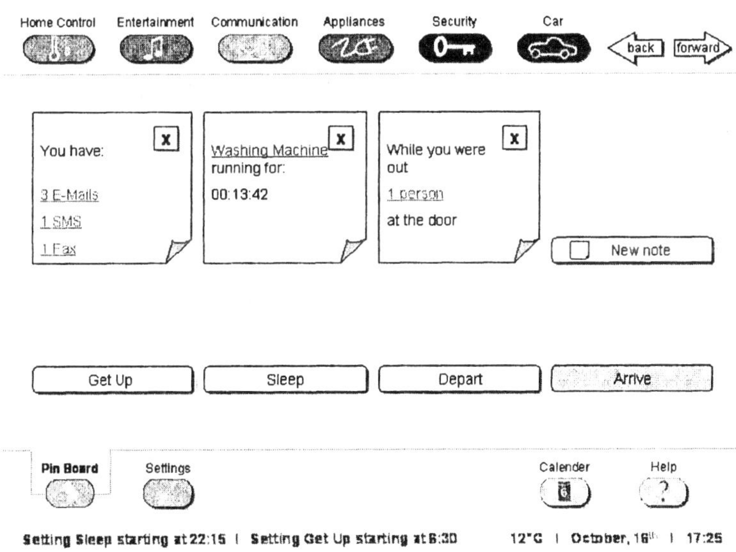

Figure 2. Touch Screen Prototype – Pin Board (starting screen)

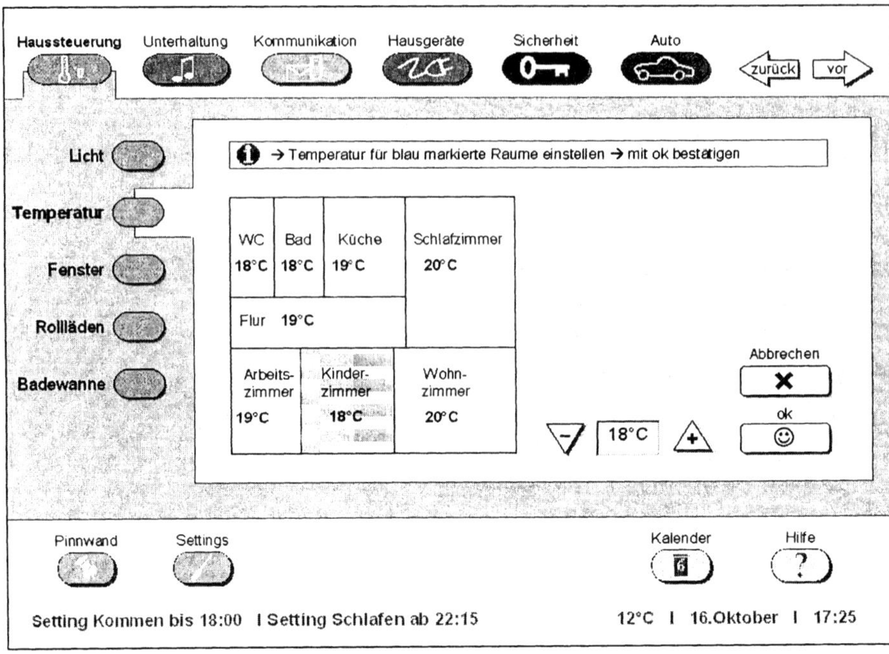

Figure 3. Touch Screen Prototype – Heating control screen with instruction, floor plan, parameter change and cancel/ok button

3.2 Adaptation to device

3.2.1 User interface concept Touch Screen

The touch screen display user interface was developed first. With this development an iterative optimization of the initial information structure and wording was achieved. The touch screen display user interface underwent a layout review, and two usability tests (Ringbauer et al., 2003). A third usability test was made with the final version to test it with elderly users (Wiebelt & Ringbauer, 2004).

The prototype is displayed in fig. 2 and fig. 3. You can see a pin board screen serving as a home screen for the smart home control. The pin board is the central place for information exchange between household members and between people and system. The settings (combined functions) are situated on that main screen to enable easy and quick access (e.g. when leaving the house).

The main navigation on the top is organized in functionally oriented tabs. Orientation is supported by visual clues like colors and icons.

The navigation steps to take a control action are:

1. Select application domain (particular tab in the top navigation, e.g. home control)
2. Select specific application (e.g. heating)
3. Optionally: Read the instruction
4. Optionally: Judge on the basis of the overview provided if a control action should be taken
5. Select a room for parameter adjustment (e.g. children room)
6. Adjust parameter (e.g. from 18 ° C to 20 ° C)
7. Confirm or cancel the action

3.2.2 User interface concept PDA

The PDA user interface prototype was implemented in flash on a Compaq iPaq PDA. In our prototype we did not include the hard keys of the PDA in the interaction concept because we assume that the PDA is not only used for home control but also as a personal digital organizer. Therefore the default hard key functions "address book", "calendar", "tasks" and "memos" were not overruled by smart home functions. The use hard keys to control smart home appliances, e.g. one-click access to the pin board screen was not in the scope of this study but should be subject of future research. The keyboard symbol in the prototype (see fig. 4 and fig. 11) simulated the inclusion of the PDA on-screen keyboard in the interaction concept (e.g. to write a notice).

A PDA-specific adaptation was made to the contents provided. The PDA offers less functions and services than the touch screen display does: there is no opportunity to build a complex setting (e.g. to define what the smart home system should do when the family is going asleep) or to change a complex setting but to activate it (e.g. press the icon "Sleep/Schlafen" in fig. 4 resp. fig. 11). I.e. the semantic consistency Denis & Karsenty (2004) claim for inter-usability of multi-device-systems is not given - touch screen display and PDA are complementary devices (Denis & Karsenty, 2004). On the other hand a PDA user interface gets clustered and complex quite easily when to many functions are provided. Here the inter-usability of the multi-device system has to be balanced to the usability of the respective device.

Having defined the PDA as an additional user interface to the main user interface touch screen display, the PDA does not have to be used by everyone in the household but is probably used together with the main device for a household member owning a PDA and needing remote access.

For the PDA user interface concept this means that we have to face device transitions and we have expectations on the one device because of the other one. In addition a PDA is normally used as personal device for other applications and services than house control as well. So we have an experience background of the person using the PDA with these applications (see section 2).

Several user interface decisions on the balance of inter-usability and handheld usability had to be taken for the PDA concept:

a) Is it more important to have a *perceptual consistency* between touch screen display and PDA concept or to allow a combination of icon and textual description of the functions in the top navigation?
b) Is it more convenient to work with floor plans like in the touch screen prototype or more efficient to work with lists and dialogues on the small screen size of the PDA?
c) PDAs work mainly with the concept of auto-save (Weiss 2002) in contrast to a dialogue pattern with ok/cancel. Because of better results in the usability test the dialogue concept was implemented for the touch screen display instead of the auto-save concept. What should be adopted for the PDA?
d) Do the preferences depend on experiences made with the touch screen device or with other mobile devices before?

The low-fidelity prototype to investigate these questions is shown in figure 4 - figure 9.

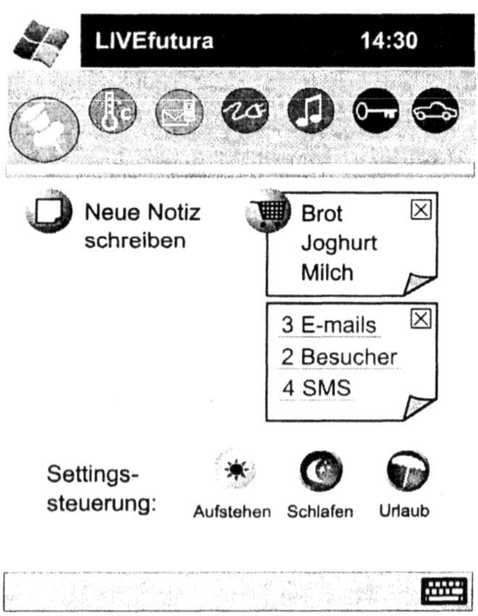

Figure 4. PDA lo-fi prototype - top navigation with icons in a row

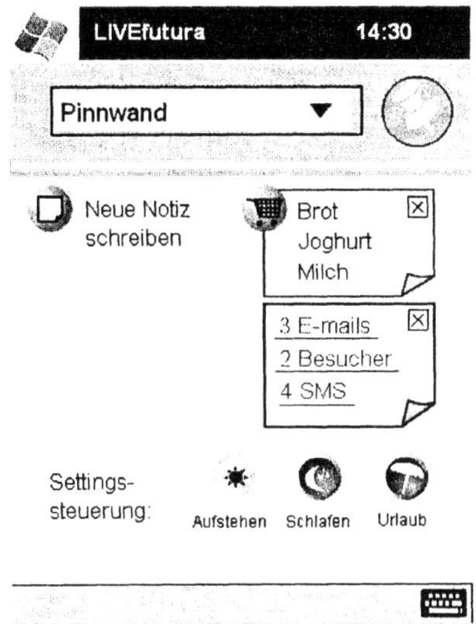

Figure 5. PDA lo-fi prototype - top navigation with icon and additional label with drop-down box navigation

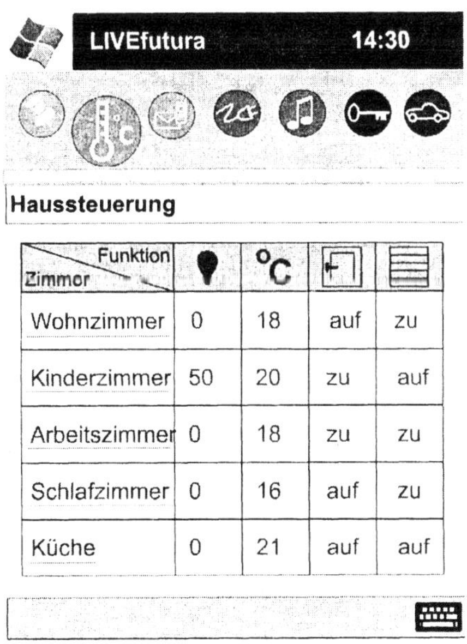

Figure 6. PDA lo-fi prototype - list-type overview on the application domain home control; parameter change is achieved via choice of room or function

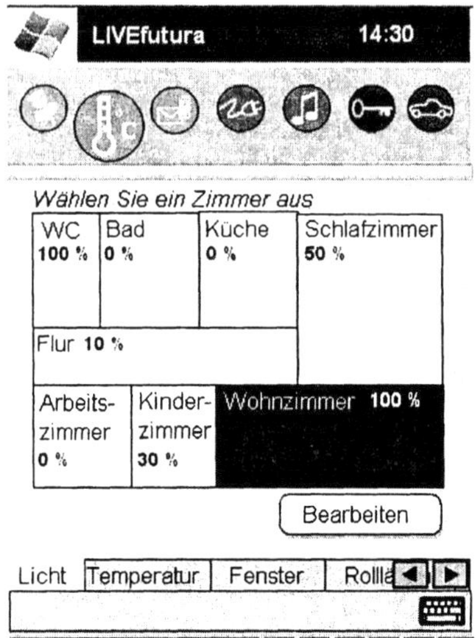

Figure 7. PDA lo-fi prototype - floor plan navigation with one function on a single tab

Figure 8. : PDA lo-fi prototype - auto-save concept in the style of the Pocket PC

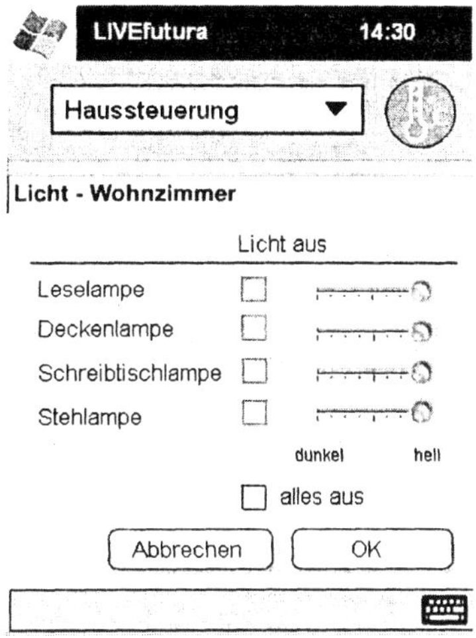

Figure 9. PDA lo-fi prototype - dialogue concept with "cancel" and "ok"

3.3 Results of the usability tests with the PDA prototypes

The usability test was done in august 2003 with ten test persons from whom four had participated in the usability test with the touch screen display prototype as well (this was done to get an idea of an answer to question d). Half of the test persons used a PDA in everyday life and gender was balanced.

a) *Perceptual consistency vs. Icon plus Wording*: The answer to this question is very clear: the versions with the icons in a row were highly preferred to the drop-down box. This stated also for the group who did not know the touch screen prototype. Some persons said, they would like some additional wording to the icon but all said that they would get used to the icon without wording as well because of the colour. There was no usability problem with either version but test persons said that the version with the icons in a row would support the transparency of the application as a permanent overview of functions.

b) *Floor plans vs. lists and dialogues*: The answer to this question is not obvious. The overview on the status of the house concerning a specific application domain was better supported by the list - some test persons did not note being on a particular tab (e.g. light or heating) at all (again no noticeable difference between touch screen and naïve group). On the other hand people were annoyed by having to step through the dialogues. They wanted to change the parameters in the list directly.

c) *Auto-save vs. dialogue confirmation* : We saw a preference for the confirmation buttons by most of the test participants - it seemed to be the more secure type of interaction. One reason was that the Pocket PC-style "ok" to close the parameter-change-window was overlooked by most test participants and that "ok" without the opportunity to "cancel" seems to cause discomfort. On the other hand the users being more experienced with mobile devices said one would easily get used to that. We also saw a learning effect concerning the use of the "ok" in the upper right corner in the course of the test.

d) *Previous user experience*: We did not see a strong effect of either having tested the touch screen display prototype before or of using a PDA in everyday life.

3.4 Acceptance of devices

After the usability test with the PDA we had an interview about which device would be the best for interacting with an integrated home environment.

Seven out of ten people said a PDA would be the best control device followed by a stationary touch screen display (five out of ten). Even four out of the five people who also tested the touch screen display preferred the PDA[11]. The main reasons were that a PDA was said to be cute in size and provides an easier feedback than a remote control without integrated display (e.g. the standard remote control of the TV). Negative comments to the PDA use were that having to use a stylus for every action is uncomfortable (as described in section 3.2.2 there was no hard key interaction for the prototype).

This might be an important point: happily using a PDA as an input device in a usability test does not guarantee it being a suitable input device in every day use in the home. Living in the smart home test environment *INHAUS* in Duisburg we found out that a PDA is not suitable for many contexts of living like sleepiness, wet hands etc. (Ringbauer & Hofvenschiöld, 2004).

4. REDESIGN OF PDA PROTOTYPE

On the basis of the test results the PDA user interface concept was redesigned and extended. We streamlined the PDA on more efficient task completion. The main navigation at the top with the icons in a row was kept

[11] The test device for he touch screen display was a big web pad. The touch sensitivity of the web pad was not optimal. So we could have an effect on the preference of device for this reason.

but the style of the application was harmonized within the application and to other applications used on a PDA (e.g. standard tabs in contrast to the button-like tabs we used in the touch screen display to avoid a "desktop-style").

The note on the pin board of the low-fidelity prototype has been limited to the number of three. Working with the notes in the usability test was straightforward. Therefore for the redesign we decided to display a list with icons but to keep the note-metaphor when reading a note.

To face the display limitations of the PDA without forcing the user to take too many steps to complete a task we decided to add specific navigation for the refined PDA user interface concept. E.g. there are "previous/next note" buttons for the PDA when a note is read to compensate for the note overview on the touch screen display.

The pin board screen with the list of notes is shown in fig. 10. You can see an open note in fig. 11. The buttons "vorherige/nächste" translates to "previous/next".

Figure 10. PDA user interface - pin board screen with a list of notes and the most important settings (waking up, sleep, holiday)

As status information displayed in a list (e.g. the status information of lightning, heating, windows, shutter) was appreciated in the test, the refined PDA user interface concept offers list-type overviews with the opportunity to directly change parameters on the same screen on the first tab. This supports the important use case of remote access and respects the less space a PDA offers. On the other hand we decided to redundantly provide floor plan navigation on the other tabs to support a visually easy to grasp overview on a single application, e.g. lightning. The interaction opportunity to zoom in a specific room from a floor plan could be a future starting point for a better context-driven task support, given the user is situated in that room.

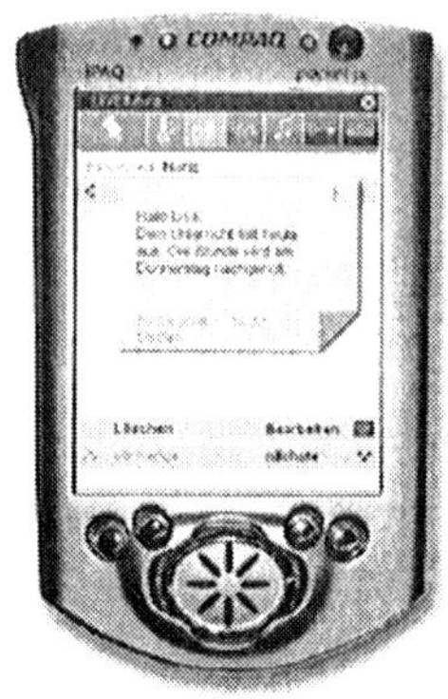

Figure 11. PDA user interface - open note with "previous/next" buttons on the bottom

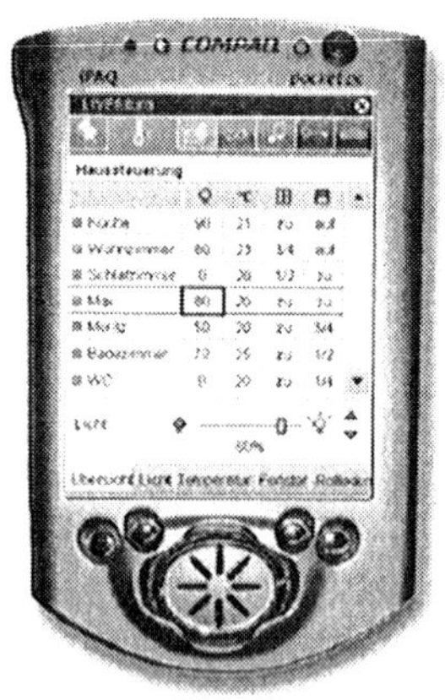

Figure 12. PDA user interface - list-type overview on home control services; lightning in Max′ room selected, direct change of the parameter setting possible

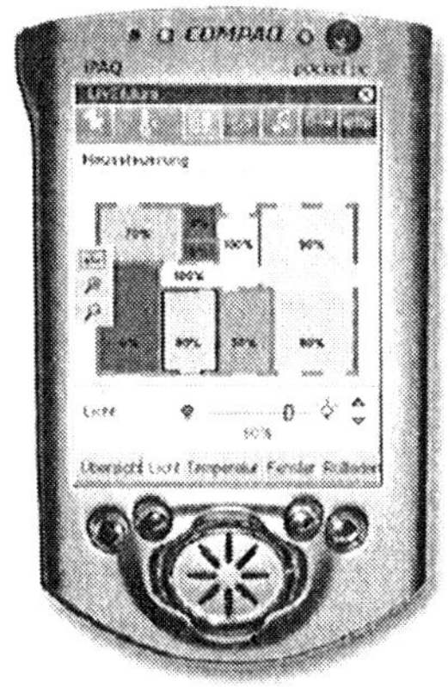

Figure 13. PDA user interface - floor plan with an overview on lightning

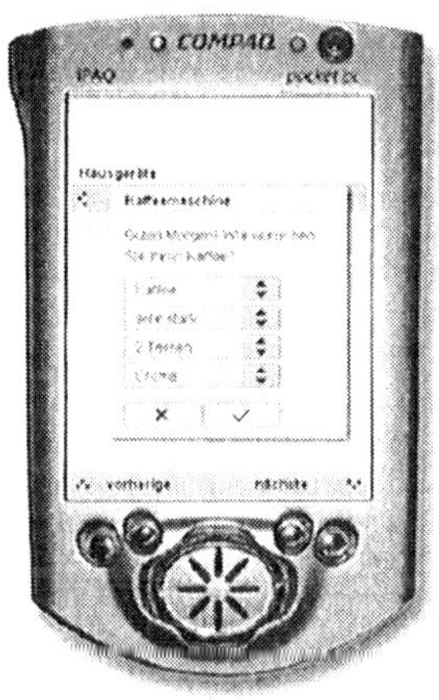

Figure 14. DA user interface - coffee machine screen with dialogue to order coffee; "cancel" and "ok" provided

We found a compromise to deal with the findings that the dialogue-style interaction concept that requires an "ok" or "cancel" to terminate each action was preferred but an efficient task completion was also claimed (e.g. an adjustment of temperature directly in the list-table would have been appreciated). The refined PDA user interface concept forces "ok" or "cancel" only when an "action" is performed in contrast to a "parameter change". E.g. editing of a note or ordering a cup of coffee leads in a dialogue and requires a confirmation whereas the change of temperature in a room or closing of a window works with auto-save. To avoid usability problems arising from the mixture of auto-save and dialogue-style we provide clearly visible and real-time feedback on the parameter change.

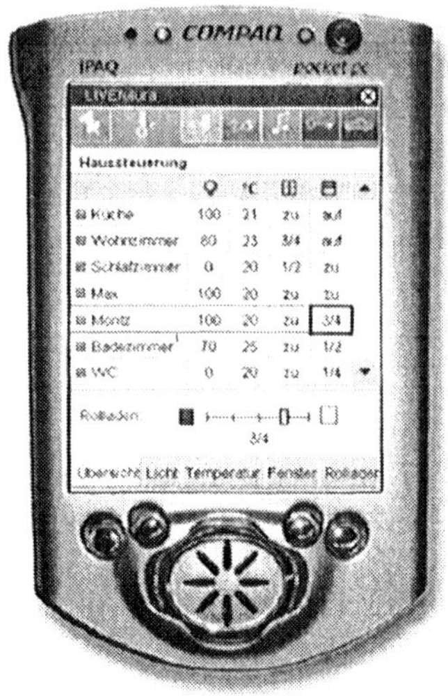

Figure 15. PDA user interface - shutter control in the overview list; auto-save concept

5. CONCLUSION

We suggest that in future the interaction with a smart home environment is accomplished with a family of devices, depending on task, situation and actual user. These devices will be complementary devices (Denis & Karsenty 2004) in the sense of being tailored to respective user and usage scenarios. One important device will still be some manual control as this is a type of interaction people still feel secure with and want to keep (Ringbauer et al., 2003).

What turned out to be important for a usable multi-device user interface is a good balance of inter-usability and usability of the respective device. Here consistency-issues have to be balanced with task effectiveness. In our study some principles of inter-device consistency were overruled by device-specific requirements.

Perceptual consistency: we suggest that perceptual consistency is important for a common branding and from a usability side concerning the recognition of main interface elements and principles. The type of interaction element should be the same but the visual appearance can be adapted to what is expected on the respective device.

Lexical consistency: if there are labels, they should be the same across devices. So requirements of all possible devices or interaction means for the whole multiple user interface has to be taken into account from the beginning. This might be even more complex having a voice user interface to develop.

Syntactical consistency: if there is no reason to change the order of steps to take to accomplish a task on different devices, keep it. Otherwise it is important to keep transparency of the interface to make clear why steps are different. E.g. one could provide a personal mobile device to control the smart home environment as some kind of "personal remote control in the house". Then the user interface should take into account that it is more likely that the user wants to control the lights in the living room being in the living room than controlling the lights in the cellar.

Semantic consistency: services provided should only be the same if that is suitable for the scenario the device is meant for. One rule should be that there is no function grouped into another category for different devices (e.g. if there are the application domains "communication" and "security" the cameras in the house with the option to leave a message are either grouped in communication or in security - this does not depend on the device.

A smart home interface does not have to be able to show the actions of the actual user but the actions of the system (e.g. rule-based control the heating system) and the actions of others users of the house.

6. LIMITATIONS OF THE STUDY

In our study we collected data on the usability of multi-device systems mainly in the laboratory on the basis of our requirements analysis in the field. For the future it is important to extend the studies on multi-device systems to real living and testing environments to acquire more knowledge on context and habituation effects. Especially the systematic investigation of device changes and on user characteristics determining device use will be of interest.

7. FUTURE CHALLENGES

On the technical side the provision of task continuity on all devices in a system will be a future challenge. On the user interface side, one interesting challenge will be how to profit from adaptation and context-awareness without causing the feeling of loosing control over the system. Another challenge will be to find a balance between automated generation of interfaces for different devices and manual tuning to enhance usability. Concerning the development process of a multiple user interface there will be a challenge to plan for interaction options that are not available yet and to manage the multi-device development: nowadays the companies are often organisationally structured on the basis of the service types and devices they offer - they all have to work together and synchronize their activities on multiple channels for a usable multi-device user interface.

REFERENCES

Ahmend Seffah & Homa Javahery (2004): Multiple User Interfaces: Cross-Platform Applications and Context-Aware Interfaces. In: Ahmend Seffah & Homa Javahery (eds.): Multiple User Interfaces: Cross-Platform Applications and Context-Aware Interfaces. Chichester: John Wiley & Sons.

Carrol, J. M. (2000). Making Use. Scenario-Based Design of Human-Computer Interactions. Cambridge, MA: The MIT Press.

Denis, C. & Karsenty, L. (2004): Inter-Usability of Multi-Device Systems - A Conceptual Framework. In: Ahmend Seffah & Homa Javahery (eds.): Multiple User Interfaces: Cross-Platform Applications and Context-Aware Interfaces. Chichester: John Wiley & Sons.

Harper, R. & Shatwell, B. (2003). Paper-mail in the Home of the 21st Century. In: Harper, R (ed.). Inside the smart home. London: Springer.

ISO (1996). ISO 9241-10: Ergonomic requirements for office work with visual display terminals (VDTs) -- Part 10: Dialogue principles. Geneva: ISO

ISO (1999). ISO 13407: Benutzer-orientierte Gestaltung interaktiver Systeme. Berlin: Beuth.

Kirwan, B. and Ainsworth, L.K (1992). A Guide to Task Analysis. London and New York: Taylor and Francis.

Klamer, L. (ed.) (2001): Screenfridge Rapport. TDC Tele Danmark og e2 Home.

Lynch, P. J. & Horton, S. (2001) Web Style Guide: Basic Design Principles for Creating Web Sites, Second Edition. Yale University: Yale University Center for advances instructional media.

Meyer, S. (2002). Was will der Kunde - Was nutzt der Kunde? Trends, Wünsche, Vorbehalte. Convention documentation of the e/home 2002, Berlin.

Microsoft Corporation (2001). Microsoft Windows User Experience. Official Guidelines for User Interface Developers and Designers. Microsoft Press.

palmsource™. Palm OS® User Interface Guidelines. Retrieved November 12, 2004 from http://www.palmos.com/dev/support/docs/uiguidelines.pdf

Ringbauer, B. & Hofvenschiöld, E. (2004). „Was macht es denn jetzt?“ - Emotionale Faktoren bei der Akzeptanz von Smart Home Lösungen. In: M. Hassenzahl und M. Peissner (Hrsg.): Usability Professsionals 2004, Tagungsband des 2. Workshops des German Chapters der UPA, S. 87-89. Stuttgart: German Chapter der Usability Professionals Association.

Ringbauer, B., Heidmann, F. & Biesterfeldt, J. (2003). When a House Controls its Master - Universal Design for Smart Living Environments. In: J. Jacko und C. Stephanidis (Hrsg.): Proceedings of 10th Int. Conf. on Human-Computer Interaction, (HCI International 2003), Vol. 1, Crete, Greece, Juni 2003, S. 1228 - 1232

Stephanidis, C. & Savidis, A. (2003). Unified User Interface Development. In: Julie A. Jacko & Andrew Sears: The Human-Computer Interaction Handbook. London: Lawrence & Erlbaum Associates.

Sun Microsystems (2001). Java™ look & feel design guidelines, second edition. Boston: Addison-Wesley.

Usability Net. Card Sorting. Retrieved February 5, 2003, from www.usabilitynet.org/tools/cardsorting.htm

Weiss, S. (2002). Handheld Usability. Chichester: John Wiley & Sons.

Wiebelt, F. & Ringbauer, B. (2004). Universelle Benutzbarkeit der LIVEfutura Steuerung mit Schwerpunktsetzung auf Senioren. In: M. Hassenzahl und M. Peissner (Hrsg.): Usability Professsionals 2004, Tagungsband des 2. Workshops des German Chapters der UPA, S. 19-22. Stuttgart: German Chapter der Usability Professionals Association.

Windows Mobile-Based Pocket PC User Interface Guidelines. Pocket PC User Interface Guidelines. Retrieved November 12, 2004 from http://msdn.microsoft.com/library/default.asp?url=/library/en-us/ui_guide_ppc/html/PPCUser_Interface_Guidelines_SKLK.asp

LEARNING FROM LOSEABLES

An exercise in critical reflection

Darren J Reed
University of York, U.K.

Abstract: There are a number of themes and descriptive categorisations of devices in HCI that act as positive common places. These include portable, wearable, and ubiquitous devices. In the this paper the category 'loseables', which includes misplaceables, forgetables and stealables, is offered as an alternative formulation of the self same devices. In a recent keynote address David Benyon proposed that HCI practitioners could utilise their own craft skill and tacit knowledge as users of devices to generate questions about design. He also suggested that there was a place for constructive criticism in relation to design that entails a role similar to that of the literary critic. The critical reflexivity method presented in this paper draws inspiration from these comments, and in combination with Philip Agre's idea of HCI as a 'critical technical practice', offers an exercise in reflective HCI. It explores the essentially contested meaning inherent in devices and, in so doing, presents not only commentary on design itself - along with the simultaneously constructive and constrictive nature of such terms - but also, and more importantly, generates questions, insights and suggestions for design.

Key words: devices, categorization, HCI.

1. INTRODUCTION

This paper started with a witticism: having discovered that my PDA had spent the night in a bush outside my house, I commented to a colleague that it wasn't in fact a portable device, ... it was a loseable[12]. As I told the story to more people, they replied with their own examples of the ways that they had re-seen the technology around them. As I thought about it, I realised that the act of deliberately redefining the character of the device in this way lead to interesting opinions, questions and insights. What was more, such efforts revealed a prevailing hegemony with HCI: a set of taken for granted terms that serve as common-places that act to close down discussion about a device's meaning.

1.1 Learning from loseables

The question became how this insight could be used productively. For example we might ask how we turn a loseable into a findable - the key ring that beeps when whistled at comes to mind here. It was realised that these alternative categorisations, and their application, open up opportunities for reflection. Critical reflexivity is a first pass at inscribing and advocating a method of renaming, and this paper is a first gathering of what might be some of the method's outcomes and insights. These include a means to work up criticisms of particular categories and designs, the (re-)discovery of the contested nature of a device's meaning and what that might mean for 'design-in-use', and the revelation of examples of experience-through-dialogue with a device.

Underlying the method and its application is the issue of reflexivity.

2. CONCEPTUAL UNDERPINNING

2.1 Reflexivity

While 'reflexivity appears in many different costumes on many different stages' [2] (p. 6). it is generally understood that reflexivity is 'the capacity of language and of thought – of any system of signification – to turn or bend back upon itself, to become an object to itself, and to refer to itself' (p. 2). To Garfinkel, reflexivity is the primordial human capacity [5] and the route through which the lived world is made meaningful. According to the sociologist Alfred Schutz, 'lived experience acquires meaning and is understood only reflexively' [12] (p. 42).

12[12] I have maintained the (mis) spelling of losable as loseable because of the more obvious combination of loss and ability. The addition of 'able' at the end of words seems to me to be an interesting technique for foregrounding the experience of the active agent or user.

Ashmore [1] draws out a number of distinguishable definitions of reflexivity, that equate to *reflection, constitution* and *mutual elaboration* [11]. 'R-reference' or reflection is the benign gaze of awareness (self- or other-); 'R-constitutive' acknowledges the constructive influences of looking; and 'R-circularity' conveys the back and forth relationship between object and representation that can be seen as a form of dialogue between a thing and its description - a continual and ongoing interaction that Mehan and Woods [8] call the 'hermeneutic spiral' (For a fuller discussion of all these ideas see [11]). Understanding reflexivity as 'reflection', 'constitution' and 'dialogue' is the basis of the critical reflexive method.

2.2 Critical reflexivity

What I am calling critical reflexivity then is a first pass at instantiating a method of unsettling meaning that encourages reengagement with a technical object as an artefact (socially or culturally constructed). Such deliberate effort can be seen in other disciplines and methodologies that take reflexivity and foreground its existence in everyday life. For example the 'strong programme' in the sociology of scientific knowledge, uncovers the constructed nature of science by applying a policy of 'impartiality' and 'symmetry' to scientific discovery and development [1]. The programme rejects hindsight and applies equal analysis to scientific failures and successes and looks to the same explanations and causes to both sides of the success/failure dichotomy. By asking for example why we have electric instead of gas powered refrigerators (which would not have the annoying hum), the approach reveals how social processes are the core of so-called objective reasoning.

The strong program is an exercise in irony, in that by enforcing equanimity that doesn't naturally exist in the world insight can be gleaned. In the earlier work of the author the ironic reflexive move was used to make comments about the internal discussion and relationships within ethnomethodology [11]. By creating an approach called pragmatic, ethnomethodology that embodied the 'ideal', the author set up a discussion and critique of extreme views within ethnomethodology, and showed how these ran contra to the disciplines foundational and fundamental commitments to the study of social life.

Already HCI has moved from its quantitative roots in cognitive psychology and computer science towards the qualitative insights of sociolology, ethnography and ethnomethodology. In reaction to an interest in studying technology in domestic settings, there is a growing openness to methodologies from such approaches as critical cultural studies, literary studies, and art and design. With these methods comes a move away from experimentation, inductive reasoning and incremental development toward single case arguments, deductive reasoning and individual creativity.

A practitioner of such approaches is Don Norman who is well known for providing his own individual reflections on everyday devices [9]. In a similar vein, David Benyon, in a keynote address at ECCE-12 (European Conference on Cognitive Ergonomics), proposed that HCI could utilise their own craft skills and tacit knowledge as users of devices to generate a critical voice. The idea that HCI practitioners should employ 'reflection-in-action' was the theme of a recent workshop at CHI2004 (Computer Human Interaction). Based upon the work of Donald Schön, we can see this as a move toward art practice and away from rationalised design. The growing interest in art approaches has been called the 'aesthetic turn' by Wright and McCarthy [14] (p. ix)

One way that the critical voice has already been heard in HCI is in the work of Philip Agre. In a discussion of the underlying features of AI research he suggested a form of critical self-reflection or 'critical technical practice' 'in which technology development can be not only an end in itself, but also a means to reflect on the assumptions and attitudes that underpin our [HCI's] ideas about technology and humanity" [10].

Garfinkel talks about making trouble as 'aids to a sluggish imagination': 'procedurally it is my preference to start with familiar scenes and ask what can be done to make trouble' [5] (p. 37). By doing this, layers of sense making can be revealed and engaged with. Critical reflexivity makes trouble by suggesting a feasible alternative categorisation of portable devices. This trouble making reveals design issues and insights. In a similar effort, Bell et al talk to the analytic strategy of 'defamiliarization' in the home [3]. Common-place things are revealed as meaningful and meaning is revealed as flexible fluid and contested.

3. FROM PORTABLES TO LOSEABLES

I offer an alternative way to think about portable devices.

3.1 Lose-ables

There are a number of things in our life that we are always losing: keys are not in the door, the remote control is no longer on the coffee table, and the spectacles are no longer on the bedside cabinet where you left them the night before. We have developed ways of keeping from losing these things, or at least finding them once they are lost. There's the spectacle string, the key ring that beeps at you when you whistle, and when all else fails there's backtracking – searching everywhere that you remember being in the last half hour/half day or half year, depending on your last memory of the thing (although this often changes as your options diminish).

Whether this is put down to forgetfulness, or diminishing mental capacities, losing things bugs us. We are brought to a moment of anxiety as we search the same places over and over again. The underlying commitment is that we shouldn't lose them. There are some things we simply never lose; strangely it is normally the very things that are important to us, that we use everyday, that go missing. When this happens we should call these devices what they are, they are loseables and it's their fault that they are lost. As I spoke to colleagues about my crusade of redefinition, I started to collect instances of loseables and developed a set of subcategories.

The categorisation loseables is populated by the sub categories 'misplaceables', 'stealables' and 'tidyables' (amongst others) and I suggest this categorization be used in place of the single term portable (although I could have picked a number of others). To convey these I will draw on my and other's experiences.

3.2 Misplaceables

The primordial loseable is the misplaceable. A misplaceable is a thing that it is possible to put in the wrong place, either by deliberate action (e.g., we put it there) or by accident (e.g., it fell there). An example I will draw on a number of times is my brand new PDA. It is smaller than my earlier device and it does so much more. It is more powerful, has more functions, and is definitely more useful; yet I keep losing it. I have left it at work, when I should have brought it home (and visa versa); I have left it on a chair at a workshop (pointed out to me, thank goodness, as I was leaving the room).

In so many ways this is the perfect device. It has a myriad of functions - that I actually use. Aesthetically it is pleasing; and importantly it has that wow feature that generates conversation in the hallway of the computer department. It is dear to me; yet I still keep losing it.

It isn't just that it is small, or that the case is black and blends in. The device is such a central part of my working life that I take it everywhere. It is almost that its immersion in my life means I can forget it. Like a well-chewed pen it has moments of perceptual invisibility.

3.3 Stealables

Things go missing; when they do we are forced to ask why. Could it be that we haven't misplaced it, but instead someone has taken it? They have been either moved out of the way (tidied – see later) or deliberately removed.

An example told to me by a colleague was the concern he felt about his IPod device. An IPod plays digital music. It is small (about 6cm by 10 and 2 cm deep) and is considered to be highly stylish having a white front and shiny chrome back. It is unique in having a bright white set of ear-bud headphones. The problem with these headphones is that because they are highly visible, they advertise what is on the end of them. The aesthetic qualities not only make the devise desirable to the consumer, they also make the wearer vulnerable to the crook. Interestingly the colleague who told me the story has since started to wear an alternative set of black headphones because he has lost the white ones. If we were in a psychological mood, we might make a connection between these two facts. The IPod is also a device that sits in a cradle on the desktop. Its appearance in combination with its accessibility makes it attractive to the thief and its size makes slipping it into a pocket very easy. It is the perfect stealable.

As the number of small, stylish and expensive devices proliferate, we have more and more stealables in our possession.

3.4 Tidyables

My final subcategory again draws on my experience of my PDA: I once left it at work and someone tidied it away for me: they squeezed it between things on my filing cabinet, obviously concerned that it would become a stealable. Portable devices are not always being carried around; at some point in the day they are at rest. As our homes and places of work fill up with a number of portable devices there is the greater likelihood that they need places, and that these places will not always be consensual. An example provided by a colleague is that of remote controls and power leads which in his experience tend to congregate. We have a number of remote controls and there are three main ways of dealing with them. The first involves placing them near to where they are used, typically on the coffee table. The second place, probably in reaction to the ugliness of a pile of remotes on the coffee table, is to put them next to the device for which they are designed. However instead of individual remotes next to individual devices, they tend to be put next to the larger device, typically the television. Peculiarly in my own household this congregation can also include remote controls for devices in other rooms. The other example was the power lead. Again these are multiplying and we have a choice: either we can leave them with the device or 'store' them together in a draw, say. Unfortunately the problem then becomes sorting out which lead goes with which device, on each and every occasion, and draw stores other lead, differentiating between different types of leads (see later example of 'fryables').

4. TRANSITIONAL MOMENTS IN MEANING

A colleague in the computer science department told the story of when his portable camera became a loseable camera. He wished to take the camera with him on a weekend away, but couldn't find it. He searched at work, searched at home, and searched his bag four or five times. In these moments, while he knew the camera was a loseable, he didn't know whether it was a misplaceable, a stealable, a tidyable or something else. In fact, he remarked, he would have been happier if he knew it had been stolen because at least he would have known where it was (and what it was). I have my own example of a transitional moment, one that lasted over 24 hours. I once lost my PDA and searched my house for an hour. I then went to work late at night and searched there. The next day I searched the path from my car to my front door and there it was; it had fallen out of my bag the night before and spent the night in a bush. Like my colleague during this time the device moved from being a misplaceable to being a stealable and back again a number of times. So a loseable has a undefined nature up to the point where it is found - its meaning is up for grabs. This indeterminacy caused a good deal of anxiety.

4.1 Deliberate transitions: turning a loseable into an attachable

As I mentioned at the beginning of this paper, we are active in dealing with loseables. Another colleague tells the story of how he turned his loseable PDA into an 'attachable' by purchasing a pouch, which hooks onto his belt. His PDA is now carried around with him wherever he goes. He has deliberately redefined his loseable device. What this speaks to is deliberate transitions of meaning, or what McCarthy and Wright call 'appropriation'. Another way to say this is that it is an example of 'design-in-use', which includes other activities such as augmentation and personalisation, and which speaks to the active part users of technology play in defining devices in different ways.

All of these thoughts and insights put me in mind of how useful it was when I redefined things as loseables. Calling my PDA portable simply didn't do it justice and certainly did not capture my daily experience of it. By playing with its meaning by deliberately introducing a transition of meaning I was able to see the device anew and almost instantly make suggestions about how it might be changed.

5. DISCUSSION – OUT-COMES AND IN-SIGHTS

Critical reflexivity then is an analytic version of these everyday activities of transition. However by targeting our appropriation and ascriptions we can unsettle taken for granted acceptance of the 'designed in' features of a device, reflect on its meaning and generate new design ideas.

However the deliberate act of re-defining a device also points to the reflexive nature of all naming, and the action of creating alternative meaning is an example of a dialogue with the device, a dialogue that mediates our experience of it as a particular thing (or alternatively shows how our experience of a thing works to define it and categorise it).

5.1 Design insight (R-reflection)

There are a number of design insights to be gleaned from our exercise in critical reflexivity. Firstly, of course, that a device can be defined as a loseable speaks negatively to its design. By deliberately redefining a portable as a loseable we become inclined to ask how we stop it from being a loseable. The earlier example of turning a loseable into an attachable is one practical example of adaptation or augmentation. But the question becomes do we want attachable devices? This after all would be a wearable. Do we want to wear all our devices? The wrist-watch is a case in point; and we could suggest that this has lead to (or is the result of) an increased emphasis on time and efficiency in society. Do we want to become so accountable to our personal digital assistants for example as we are becoming accountable to the mobile telephone? Do we want TV screens in our spectacles, ID card in our forearms, and the like? Design, then, at this level is about making decision about what we want technology to mean for us in the future.

5.2 Categorisation and meaning as contested (R-constitutive)

Meaning is socially constructed: in the laboratory, in the home, in the street, but perhaps more interestingly in *research insights* such as HCI. That it is possible to find an alternative categorisation speaks to the constructed nature of the meaning of a device and sensitizes us to the idea that a single device has multiple possible meanings. By contrasting portables with loseables we *re-introduced the contested nature of the meaning of (in this case) technological devices.* A practical example where multiple meanings are prevalent is the home, and we are becoming more sensitive to instances of when technology has different meanings for different family members. A nice example was given to me by a colleague who plays digital music through his computer at home. When he doesn't use headphones, he prefers to attach a set of portable speakers so that he can move to a private space and not annoy others with his music.

One interesting insight is that there are everyday moments of meaning transition, in which how a device is perceived is in flux and open to meaning ascription. In the case above it is seen in the moments between recognising a device as falling into the broad category of loseable, and when it is specified as a misplaceable or stealable. These are natural moments of flux of constitutive reflexivity. They speak to times of uncertainty and anxiety.

In terms of our perception of technical devices, Wright and Monk have said that certain devices only become visible when they go wrong, when they no longer perform the function for which they were designed. In our case we might say that the PDA only became visible when it was a problem – when it was missing. It is interesting to compare this idea to that espoused by Weiser and the ubiquitous computer,

> "For thirty years most interface design, and most computer design, has been headed down the path of the "dramatic" machine. Its highest ideal is to make a computer so exciting, so wonderful, so interesting, that we never want to be without it. A less-travelled path I call the "invisible"; its highest ideal is to make a computer so imbedded, so fitting, so natural, that we use it without even thinking about it" [13].

Bolter and Gromala [4] say that such notions are impossible. Even if the computer was 'embedded' in space and invisible in the environment, it would still have social and cultural meaning and could never become 'transparent'. It is perhaps then cultural meaning that maintains the visibility of a device; no matter how small it will always have meaning.

More than this, devices have multiple cultural and personal meaning at the same time. Blythe's 'pastiche scenarios', - which key off common literary genres and narratives - show how an IPod can have differing and competitive meaning in the same situation at the same time. For example he creates two alternative understandings, one from a narrative built on Bridget Jones' Diary by Helen Fielding, and another on Trainspotting by Irvine Welsh. The IPod at the centre of the account is at once an object of identity and expression (in how it looks and what songs are stored on it) for Bridget, and at the same time a means to generate cash for drugs for Renton.

The idea of transition also speaks to a device having multiple meanings over time. For example, we can think a device's lifecycle and see how its meaning changes over time. An example can seen in the statement, "it's not cutting edge, it's out of date". (Perhaps at times like these we would prefer that our device becomes a loseable!).

Finally, meaning transition puts me in mind of the fact that certain meaning ascriptions, hold sway, as though in certain situations certain agents have rights over the core meaning repertoires that we can apply. We might call this a form of asymmetric ascription. I have the sense that the discipline of HCI works with a set of established and largely taken for granted categorisations. These 'common places' act as resources for amongst other things research groupings (including specialised conferences and journals), funding proposals, and decisions about what is 'in' at any give time. How good it would be, then, to deliberately unsettle some of these asymmetric ascriptions. We might then ask questions like the one inferred above about ubiquitous computers. Do we want computers everywhere, and invisible? As HCI practitioners it is appropriate that every so often we question our categorisations and their influence.

5.3 Dialogue and experience (R-circular)

As we saw earlier, Schutz felt that human experience is only possible because of reflexivity. Garfinkel says that reflexivity is essential and uninteresting, by which he means that it is so ubiquitous and there is no escape (p.8). We are continually in a relationship of reflexivity with the world. Ethnomethodology understands reflexivity as best expressed though the 'documentary method'. Garfinkel describes this notion from Mannheim in the following terms,

> '...the method consists of treating an actual appearance as 'the document of,' as 'pointing to,' as 'standing on behalf of' a presupposed pattern.... Not only is the underlying pattern derived from its individual documentary evidences, but the individual documentary evidences, in their turn, are interpreted on the basis of 'what is known' about the underlying pattern' [5] (p. 78).

This circular relationship between object and representation is ongoing and continual, and can be seen as a dialogue [8]. Critical reflexivity is a way to engage in a conversation with a device, by introducing a new interpretation of the device as document. And is therefore a method of generating and revealing dialogues with devices.

An interesting feature of these types of dialogue is that they are not one way. That is it is not simply about human interpretation of ambiguous things; the device as document completes the referential circle of meaning. Seen in notions such as Norman's affordances, and when Grint and Woolgar [6] talk about how technology 'configures the user', this circular relationship is key to our experience of a device. At times there are disjunctures in this documentary relationship when we find it difficult to reconcile the prescribed categorisation (and prescribed experience) and our own experience. The IPod is typically described as a portable device; yet its battery life requires that its portability be regularly punctuated by being tethered to a charger. Prescribed categorisation and experience are in this instance at odds. In these moments we might shout 'it's not a portable at all!'

6. SUMMING UP

There are two underlying features of the above discussion: the first is a tendency towards complexity, the second a tendency towards values. As HCI moves out of the darkness of the laboratory into the light of the social world, it is necessary to deal with reflexive processes of meaning making and the emotional and experiential capital invested in technological devices. Indeed it might seem a tall order in these circumstances to substantiate a discipline that can talk about good design. But it is only by engaging in conversation with devices and the complex meanings and values attached to them, that we can hope to reveal the important issues involved, and provide for future design.

6.1 Critical reflexivity and irony

Let us be clear: the suggestion of the alternative category 'loseables' an exercise in irony. It could be that it is useful and can be applied in different situations to improve design. That is fine. However, what I am after is a method of deliberate undermining; a means to unsettle taken for granted notions and open the way to understand our devices anew. The typography is an outcome of this effort and is not meant to stand-alone. The resulting discussion and the insight gleaned from it are the point, not the particular category itself.

Just to underline this I offer the following category of 'fryables'. During my conversations with colleagues about loseables, two instances came to light. The first involved the loseable camera mentioned above, for which there is a socket to plug in a wire to transmit the pictures to a television. The socket is shaped such that it also accepts the power lead of another device. In his relief to find the camera, my colleague inadvertently plugged in the power lead, and watched bemused as smoke emanated from its insides. He suggested the term fryable as a viable alternative description of his (now charred) camera. Another example presented itself in real time during a telephone conversation with another colleague. As we spoke the colleague started to play absent-mindedly with a paper clip, and then, obviously in reaction to the interest he felt in the conversation, he idly inserted the paper clip into the firewire port on his laptop. The expletive he uttered conveyed a transitional moment of meaning as the device fizzed and went blank.

REFERENCES

[1] Ashmore, M. (1989). The Reflexive Thesis. Chicago: Chicago University Press.
[2] Babcock, B.A. (1980). Reflexivity: Definitions and Discriminations. Semiotica, 30(1/2), 1-14.
[3] Bell, G., Blythe, M., & Sengers, P. (In Press) Making by making strange: *defamiliarization* and the design of domestic technologies.
[4] Bolter, J. D & Gromala, D. (2003) Windows and Mirrors. Interaction, Design and Digital Art, and the Myth of Transparency. London: MIT Press
[5] Garfinkel, H. (1967). Studies in Ethnomethodology. Cambridge: Polity Press.
[6] Grint, K., & Woolgar, S. (1997). Configuring the User: Inventing New Technologies. In K. Grint & S. Woolgar (Eds.), The Machine at Work. Technology Work and Organization. Cambridge: Polity Press.
[7] McCarthy, J., & Wright, P. (2004). Technology as Experience. London: MIT Press.
[8] Mehan, H., & Wood, H. (1975). The Reality of Ethonomethodology. New York and London: Wiley-Interscience Publication.
[9] Norman, D. (2004). Emotional Design. New York: Basic Books.
[10] Redmile, D. Designing for Reflective Practitioners: Sharing and Assessing Progress by Diverse Communities. CHI2004 Workshop
http://sigchi.org/chi2004/program/prog_workshops.html#ws6
[11] Reed, D.J. (2002) Observing and Quoting Newsgroup Messages: Method and Phenomenon in the Hermeneutic Spiral. Doctoral Thesis. Loughborough University.
[12] Schutz, A. (1972). The Phenomenology of the Social World. London: Heinemann Educational Books.
[13] Weiser, Mark . Web page http://www.ubiq.com/hypertext/weiser/UbiHome.html
[14] Wright, P., & Monk, A. (1991). A cost-effective evaluation method for use by designers. International Journal of Man-machine Studies, 35, 891-912.

DATA SECURITY THREATS IN THE HOME ENVIRONMENT

Tony Proctor
School of Computing and IT, University of Wolverhampton, U.K.

Abstract: The aim of this paper is to assess the relevance of data security in the home environment and to identify the nature of vulnerability. At the time of writing most data in the home is stored manually. The widespread use of the Internet, home pc's and mobile devices now means that increasingly individuals (and families) are storing "home information" electronically. It is highly likely to be a trend that continues. Information on specific data security related issues is widely available in terms of "how to protect your pc" and there are also many products that are available to assist with this. The level of actual understanding that home users have of the general data security issues and the means that they take to protect themselves are not well documented. The approach taken in this paper is to look at many of the data security threats posed to existing ICT devices and to put them in the context of systems that are currently used or may be used in future, in the home environment. The paper clearly defines that it is not concerned with the broader topics of data assurance (which would include fire, flood etc). The primary focus is the potential for malicious acts being committed by unauthorised use of the systems. Where are the threats from and what can be done to alleviate them? Are the main vulnerabilities in the connectivity, the systems, or both? Since technology now softens the "environmental" boundaries, this paper alludes to the scope of the home environment. It also discusses existing solutions to perceived threats; whether they appear to be understood and whether they appear to be working. It looks for the evidence to demonstrate the significance of data security in relation to the development of "intelligent homes". The paper then takes into consideration the directions in which technology appears to be developing in arriving at conclusions.

Key words: home security, data integrity, threats.

1. SCOPE OF THIS PAPER

A broader discussion of data assurance that could address issues such as fire, flood, system availability and resilience are outside of the scope of this paper which is concerned with the committing of malicious acts through access to technology. It is also worth noting that the increasing proliferation of technology will result in the increasing use of hardware that in itself may be of high value and hence increase the likelihood of robbery and burglary in order to steal these items. For the purposes of this paper the consequences of the theft of the data will be considered but not theft of the hardware itself. The paper will not consider the issue of hoax relating to technology. There are many web sites that describe this (a good example can be found [6]). It is not a reality and any consequences are as a result of psychology rather than process.

The scope of this paper does not allow for the discussion of the threats and preventative measures for manually held data. The degree of the perceived threat to an individual probably varies (some people may destroy utility bills, bank statements by the use of a shredder or burning, others may simply bin them). Increasingly we are being encouraged to bank on-line and pay our bills on line. This suggests that the manual storage of information will be less of an issue than electronic.

2. WHY IS DATA SECURITY AN ISSUE?

The growth of the internet and increasing use of computers in the home has raised the importance of data security in home computing. Some recent forecasts [14] indicate that 75% of UK households will have access to the internet by 2011 and that this access will be by a variety of different devices.

Fifteen to twenty years ago data security was little more than making sure that systems had adequate backups. It was also a technical activity performed exclusively by System Administrators and not a personal activity undertaken by a non technical person in the confines of their own home.

In recent times we have encountered many attempts to compromise computer systems; from writers of virus software to what might be termed cyber terrorists. Over time, the profile of the hacker is changing from a realm of "young geeks" to a more mature IT Professional using their skills to engage in criminal activity. Technology now enables traditional crimes to be committed more easily and is also facilitating new crimes.

Hackers are able to exploit the inherent weaknesses in operating systems. The innumerable combinations of hardware and software means that it is not possible to test new releases to "anything like destruction" and hence operating system patches are released to close the door to the "hacks" that have opened them. Modern operating systems have many millions of lines of code. This increasing complexity of software exacerbates the problem.

From time to time the national media focuses on or exposes Information Security breaches, typically involving fraud. Often this involves corporate data, but the number of examples involving personal data is likely to increase. As the number of devices capable of storing data increases, the amount of personal data that we store electronically will also increase. Personal Digital Assistants (PDA's) and other wireless devices also present particular problems for data security.

Finally, high security and high usability are "mutually exclusive". This suggests that there is a compromise to be made between the two and this is of commercial significance to the developers and suppliers of these devices.

3. HOW IS THE "HOME INFORMATION ENVIRONMENT" CHANGING?

Just as we are more able to carry our work home with us, we are becoming increasingly able to carry some aspects of our home outside with us. Originally, devices like the telephone and home computer were strictly "home bound".

Technology is thinning the boundary between our working and domestic lives. From individuals who run a business at home to teleworkers and the many people who take work home or occasionally work at home for their convenience. ICT is the enabler, since electronic data is highly portable.

Technology that was once home-based is being used in more "creative" ways. For example, the ability to track individuals through their mobile phone (example service [13]). One of the typical functions that this could provide is for monitoring the location of family members at all times.

4. WHAT ARE THE EXISTING THREATS?

One of the main fears that individuals tend to be aware of in terms of technology crimes is fraud, "*concerns over security are stopping six million people in the UK from banking online*" [3]. Fraud is often the end product of many different, often complex computer related criminal activities. These include:

- Identity Theft. The fastest growing crime in the UK [5]. "Stealing identity to fraudulently obtain money from bank accounts, claim benefits or obtain false documentation (e.g. passports). Identity theft is used in a variety of crimes from banking frauds to people trafficking. Currently the majority of cases are accomplished via a non-technological means.
- Credit card theft. Data is normally obtained via a physical method e.g. robbery, observing a user at an ATM, installation of recording devices at ATM's rather than via the web.
- Hacking. Destruction of data, loss of service, use for third party attack (many users are unaware of this concept).
- Harassment. e.g. blackmailing and stalking
- Paedophilia. a complex threat associated with examples of a type of new crime only possible because of the Internet; such as "Grooming" and encouraging children to produce their own pornographic images, discussed by [11].

There is also a growing knowledge of and concern regarding the use of "spyware" which is one of the more recent IT security threats to be given media coverage [9].

4.1 Device or Connectivity, Where is the Threat?

Whether the threat lies in the systems or their connectivity is an issue worthy of discussion. The table in Fig. 1 helps to identify some of the more obvious systems, threats and whether they result from the systems or the environments to which they connect.

The information presented in Fig 1 identifies that the majority of threats are associated with connectivity. This is a problem that will magnify with the increasing usage of broadband and hence "always on connectivity".

Further examination shows that both devices and connectivity have vulnerabilities. Any device or medium that is capable of storing or processing data is potentially vulnerable. Where processing takes place there is a need to ensure that

- the processes are legitimate (I.E. these processes are allowed)
- change is controlled (I.E. only authorised changes are allowed to legitimate processes)

Device	Main Potential Threat(s)	Typical Vehicle for Delivery	Device (D) or Connectivity (C) Threat	Preventative Measures
Desktop PC	Altering of the way the system works to its detriment and / or the destruction of data.	Virus / Trojan / Worm Program installation or use of external storage devices	C & D A virus etc. is likely to be downloaded in some form but could be on removable media)	Regularly Updated AV / Firewall/ Spyware detection software. Care in the choice of software. Care in use of removable media
	Fraud E.G. fishing (Obtaining banking details in order to commit theft)	Web Page Rogue Program	C	Familiarity with site and procedures used by bank etc.
	Hacking; altering the way the system works to it's detriment and / or destroying / altering data by unauthorised access rather than by a virus	Cookies Rogue Programs Port Scanning / Wardialing Network / Telephone taps Inadvertent disclosure	C & D	Firewall/ Spyware detection software. Use of passwords and encryption where available. Keep login codes & passwords secret. Caution when being asked for any information relating to login codes / passwords etc. Loading of operating system patches as available. Disabling some operations (where appropriate)
Laptop	As Desktop PC	See Above		
	Portability of device	Physical Theft	D	Hide device when in transit. Use of passwords (inc. BIOS password)
PDA	Most of the threats listed for laptop also apply Less secure, as you cannot use BIOS passwords.	Most of the vehicles for delivery listed for laptop also apply	C & D	Most of the preventative measures listed for laptop also apply. However these may be less widely available for PDA's.
Phone	Unauthorised Use	Theft Virus / Trojan / Worm	C	

Figure 1. Threats, Sources & Prevention

4.2 Use of Existing Deterrents

The main categories of existing deterrents are the use of login codes & passwords, anti-virus S/W, Firewalls, encryption and physical deterrents like locks and keys.

As the major supplier of computer operating systems in the home, Microsoft offers a lot of information on security for home computing. Examples at the time of writing [10] are the top tips;

- to use an internet firewall
- get computer updates
- use up to date anti virus software

Implementation of these suggestions is often not straightforward for a computer professional or enthusiast. In theory it represents sound advice, but it would be interesting to discover how widely are they used and how well they are understood?

A preliminary survey (to provide an answer to the previous question a more rigorous study would be required) by the author from a random sample of 20 (non IT professional / non IT enthusiast) home users has indicated that;

- 5 % installed Firewall software
- 10% update their OS patches
- 75 % installed anti-virus software

Furthermore, most users in the sample were aware of the risk of viruses. Many had some idea of the purpose of a firewall (some of those talked about parental control with regard to this question). Some confusion regarding computer updates was apparent.

Even IT Professionals require discipline in order to keep up with the latest patches & releases of virus checkers, firewalls and operating system. It is also possible that installation of the latest release in order to prevent a potential security breach that hasn't yet been encountered by the user can result in the apparent failure of existing applications.

5. SECURING A DEVICE

A simple experiment was conducted by the author which demonstrated that a *reasonable* level of protection can be achieved in a short time provided that the user has a *basic* level of knowledge (words in italics may be subjective in definition and further study would require this to be defined); Firewall protection was achieved in 10 minutes, antivirus protection in 5 minutes (details in the Appendix). In both cases the user was asked questions which could be challenging for the non-technical (although defaults were provided for most choices). The software was also downloaded via a broadband link. Since many users do not have this, the time taken to download could discourage them from taking preventative measures (a factor of 5 might be considered reasonable if using a modem link).

There are a number of necessary elements for a user to acquire these "preventative measures". They need to be aware of; the need for security, where to get protection from and how to use it (the latter typically being downloading from the web and installing). There is a wealth of information about data security available from the web and there are a growing number of sites aimed at home users.

An example of a website that offers security advice for the home computer user can be found at [7]. A twelve-point protection plan for home computers is defined by the CERT Coordination Centre [12]. This provides extremely useful and practical information. Again however, technical knowledge is required. Another important point is that implementing some of the advice will result in a loss of features (e.g. disabling JavaScript / Active X). This lends further evidence to an earlier statement on the mutual exclusivity of security and usability.

6. WHAT ARE THE FUTURE THREATS?

The threats to personal electronic data will continue to grow as the dependence on / proliferation of devices that store and / or process electronic data continues to grow. There will be an increasing use of

- Smart Cards; may in future contain critical information i.e. personal medical records...what would the implications be if someone could change these?
- Embedded Devices; many may be managed & controlled by service providers. What if their systems are compromised?
- Collaborative Systems; modern cars are already using sensors and bus lines instead of switches and circuits. What are the implications if a line or an intermediate device could be controlled by external intervention?
- Decreasing usage / availability of "manual" methods for backup. Increasing reliability and availability may make users complacent.
- Loss of power; should this be considered as part of a "broader context "of security.

7. INTELLIGENT HOME CHARACTERISTICS

Across the world many countries have Smart Home projects. Some European examples include Deutsche Telecom and partners (in Germany [2]) and Tampere University of Technology (in Finland [4]). Major IT suppliers are also committing their resources, "Intel has invested $200m in companies developing hardware and software technologies for the next-generation digital home" [8].

Telemetry will play a major part; the main embedded devices in an "Intelligent Home" will be Control Systems. Good examples of this which can already be seen are Anti-Theft, E.G. Alarm, Video Camera, Sensors, Lights on / off / intensity, curtain opening & closing and audio-visual control to make it appear that the home is occupied. Human monitoring systems are also under development using intelligent software to monitor patterns of behaviour and issue alerts on exceptions. Other devices will allow access control (for example opening and unlocking the door to your home via a mobile phone or proximity device).

Many future common "household items" will also store data. This category includes Fridges, Cookers and virtually anything electronic involved in the many processes that take place in the home. One of the key considerations is which devices will talk to each other and the means by which they will achieve this.

How much control will the user have? Always on devices are always on! We can pull the cable, but what if we depend on the communication link for a different vital process? (It is no longer possible to turn off the Internet). It is also possible that our homes may act as host to third parties. There are currently some examples of wireless network providers offering discounts to homes that will host access points. This concept further complicates the data security situation.

When it is considered that the security measures taken are strong, it may place systems in a more vulnerable position when they fail; many people have got into the habit of ignoring burglar alarms and car alarms.

Chip and Pin is regarded as being more secure than signing when using a credit card. But if an unauthorised person knows the PIN then they can potentially do more damage, more easily (the theory is that this will not happen as the card readers will download the stolen card data and render a stolen card useless by writing to the chip).

Interesting discussions of some of the consequences if a malicious agent were to be able to eavesdrop on home sensors or manipulate home actuators can be found in [1]. These range in severity from ease of burglary, stalking and control of the home environment to the creation of what they term a "poltergeist virus" which turns the automated appliances in the home against the home owner.

Other than actual physical damage there is currently very little attack on hardware. Is this likely to change? Can code be written to directly interact with the physiology of devices to render their components un-usable, for example by use of fatal electrical or electrostatic charges (EMP attack). This will be more easily achieved with the move to solid-state devices.

8. CONCLUSIONS

Malicious actions caused via access to home technology are a serious factor, which could prevent the uptake and growth of this type of technology. There is potential for customers to be discouraged from buying and using mobile technology because of concerns related to data security. The extent to which this is a factor is likely to be influenced by media coverage given to incidents that will occur over time.

The more personal data that is stored, the more important the data, the more it is necessary to formally or informally perform personal data risk assessment. This involves asking?

- What data do we use and how do we use it
- What are the implications if we loose our data / if it is stolen? (To what extent could an unauthorised user achieve the activities of the previous point?)
- What data do we need to protect?
- How do we protect it?

It is unlikely that home users have a disaster recovery plan. In the event of a catastrophic system failure they are much more likely to reach for the Yellow Pages or to contact a knowledgeable friend or relative who has some technical knowledge.

So if the existing data or systems are of a sufficient importance or when they become ubiquitous, such a plan will need to be in place.

The issue of data security needs to continue to be a process of education as well as the use of security features and tools; more security does not always mean that something is more secure. It is sometimes taken for granted that a system that has more security is less likely to be violated and hence the implications if it is violated, are much greater. The main stakeholders in this "education" should be the suppliers, the government and the individual.

In terms of the homes of the future it can be seen that the increase in the processing, storage and communicating of electronic data associated with the increasing use of automation increases risk. For example, information could easily be obtained by a hacker to find out the times when the home was being "administered" remotely. Whilst there is no evidence to date, the fantasy of a remote burglary may in future become a reality.

The degree to which we become dependent on home automation and the associated applications has a major bearing on the importance and nature of security precautions. As an example, the concept of location tracking via mobile phones has been mentioned earlier in this paper. Consideration in this case would need to be given to the ease with which this "tracking" could be intercepted by a third party and what the possible implications might be.

Becoming the innocent third party in a hacking attack is a frightening but real possibility (already present in the commercial world). It is likely that there will be more use of third parties by hackers and organised IT criminals as they try to cover their tracks. This is more likely to affect organisations and industry rather than the individual but is likely to have some affect on the home user. It may be expected that in the event of such a claim by the victim, procedures (e.g. Insurance Cover) would be in place to compensate.

This paper clearly identifies that it is the connectivity that provides the major security weaknesses. In line with this, further protection will be needed to supplement the growing provision for data connectivity security currently available (e.g. virus detection and firewalls). The issue of protecting the actual devices may have less of a focus but this must not be neglected (particularly with the rate of technological advance in respect of small scale storage devices).

The paper has concentrated on loss of data or the implications of "unauthorised changing of data". As an increasing usage produces a dependence on personal ICT devices there will be an increasing importance placed on availability and continuity. This brings with it a whole host of other, related issues including operating system resilience and continuity of energy supply. Equally, the theme has been the unauthorised access to and misuse of data to engage in criminal activity.

The technology that is used to help prevent unauthorised use / access to home devices will depend upon the infrastructure(s) deployed. Whilst there are a number of connectivity proposals there is no single standard at this time. However, a reasonable assumption is that the Internet Protocol (IP) is likely to be a part of many systems. Good security requires a multi-faceted approach. The widespread use of encryption coupled with Biometric devices appears to offer considerable hope for genuine security and will be experimented with and deployed in the smart home where proven to be effective.

The current requirements for home automation require technical knowledge of some level (e.g. the use of a computer). There are many benefits to be gained from connecting the home to the internet since it allows for remote monitoring, improved control (more information is available) and ease of update. In order for home information to become ubiquitous, the level of technical knowledge required needs to reduce and this is likely to result in a dramatic increase in the number of embedded devices. These devices will be connected to the Internet and may be managed by service providers. Whilst this will provide ease of use it will also present security concerns. Mobile technology e.g. mobile phones are one of the proposed interfaces for facilitating external control. This further complicates the security perspective because it is then necessary to ensure that the communications between these devices and the home / host management system are secure.

REFERENCES

[1] Bergstrom P., Driscoll K., Kimball J. (2001) Making home automation communications secure. *Computer*, (October 2001), pp. 50 – 56.

[2] Miller, F. (2001) Information technology: wired and smart: from the fridge to the bathtub. *Fraunhofer magazine*, (2. 2002), pp. 30 – 32. Available from world wide web: http://www.inhaus-duisburg.de/en/projektbeschreibung/rojektbeschreibung.htm

[3] Greek, d. (2004) News: security breach hits another online bank. Computer active [online]. Vnu, [accessed 09 november 2004].. http://www.computeractive.co.uk/news/1159273

[4] Pirhonen, A.(2003) e.finland [online]. E.finland, [accessed 10 november 2004]. Intelligent and cosy e-home: the e.finland weblog. http://www.e.finland.fi/netcomm/news/showarticle.asp?intnwsaid=15743

[5] Halliwoll, J. (2003) Combating identity fraud. International professional conferences. IPC, [accessed 11 november 2004]. Available from world wide web: http://www.ipc-conferences.co.uk/idfraud/idfraudframe.htm

[6] Anon (2004) Hoax warnings, F-secure [online]. F-secure.com, [accessed 11 november 2004]. http://www.f-secure.com/virus-info/hoax/

[7] Anon (2004) Data security in our connected world is now about information security. Close protection [online]. [accessed 11 november 2004]. http://www.closeprotection.ws/data-security.htm

[8] Jacques, R. (2004) Intel inside the home of the future. computer active [online]. Vnu, [accessed 11 november 2004]. http://www.computeractive.co.uk/news/1151886

[9] Ward, M. (2003) The spy inside your home computer bbc [online]. [accessed 12 november 2004]. http://news.bbc.co.uk/2/hi/in_depth/sci_tech/2000/dot_life/2487651.stm

[10] Anon (2004). Security at home: protect your pc. *Microsoft.com* [online]. Microsoft, <http://www.microsoft.com/athome/security/protect/default.aspx>.[accessed 09 november 2004].

[11] Anon (2004) United kingdom threat assessment of serious and organised crime 2003: 9. Sex offences against children, including online abuse. *National criminal intelligence service* [online], [accessed 09 november 2004]

<http://www.ncis.co.uk/ukta/2003/threat08.asp>
[12] Anon (2001) Cert coordination centre: home network security. *CarnigieMmellon software engineering* [online]. Carnegie Mellon University Pittsburgh, [accessed 11 november 2004]. <http://www.cert.org/tech_tips/home_networks.html >.
[13] Dolby, J. (2003) Home: Telecomms: new mobile phone tracker pinpoints and verifies locations over all UK networks. *pressbox.co.uk* [online]. [accessed 17 december 2004]. http://www.pressbox.co.uk/detailed/10083.html
[14] Anon (2001) Short survey of published material on UK key trends 2001-2011 *UK cabinet office performance & innovation unit.* Available from world wide web: http://www.number-10.gov.uk/su/key.pdf [accessed 17 december 2004].

APPENDIX – ANTI-VIRUS AND FIREWALL SOFTWARE INSTALLED

AV Software

Installation of Sophos
File Size; 14.5 MB
Time to install; 5 minutes (including un-zipping)

Firewall

Installation of Zone Alarm 5.1 from Zone Labs.
File Size; 8.2 MB
Time to install; 10 minutes (to view the optional tutorial was an additional 6 minutes)

(please note that the choice of products was random)

REMINISCENCE VIDEO

Helping At-Home Caregivers of People with Dementia

Noriaki Kuwahara, Kazuhiro Kuwabara, Nobuji Tetsutani and Kiyoshi Yasuda

ATR Intelligent Robotics and Communication Laboratories, Kyoto, Japan. Nobuji Tetsutani, also of Department of Information and Arts, Tokyo Denki University, Saitama, Japan. Kiyoshi Yasuda, also of Department of Rehabilitation, Chiba Rosai Hospital, Chiba, Japan.

Abstract: Providing good care to people with dementia at home is becoming an important issue as the size of the elderly population increases. One of the main problems in providing such care is that it must be constantly provided without interruption, and this places a great burden on caregivers who are often family members. *Networked Interaction Therapy* is what we call the methods designed to relieve the stress of people suffering from dementia as well as that of their family members. This therapy aims to provide a system that interacts with people with dementia by utilizing various engaging stimuli. One such stimulus is a reminiscence video created from old photo albums, which is a promising way to hold a person's attention for a long time. In this paper, we first present an authoring tool to assist in the production of a reminiscence video and then discuss the effectiveness of presenting such a video to people with dementia.

Key words: Dementia, home care, therapy.

1. INTRODUCTION

As the size of the elderly population increases, the number of people with dementia inevitably increases. Providing good care to these people at home is becoming an important issue. One of the main problems in providing this care is that it must be constantly provided without interruption, and this places a great burden on caregivers, who are often family members. *Networked Interaction Therapy* is our name for the proposed methods to relieve the stress of both people with dementia (users) and their family members. These methods use a system that interacts with the users to hold their attention. The human caregivers can thus rest while the user is engaged in the system. We also assume that the user receives a positive stimulus from the system.

In this paper, we first introduce the overall concept of Networked Interaction Therapy and the issues involved in designing such a system. Then, we present a reminiscence video as a promising stimulus for people with dementia, and we propose an authoring tool to support production of the video. Finally, we discuss the effectiveness of presenting the reminiscence video to people with dementia along with the results of experiments conducted at their homes.

2. NETWORKED INTERACTION THERAPY

A system for Networked Interaction Therapy needs to get people with dementia (users) involved and keep their attention. Figure 1 illustrates an example of the home terminal for Networked Interaction Therapy. Currently, we plan to provide our service by using a large-screen TV and a set-top box that controls network communication along with a camera and microphone for monitoring the user's status. We are also considering using an air cannon and a vibrator in the sofa to activate the olfactory [1] and haptic senses [2] in addition to the audio and visual stimuli.

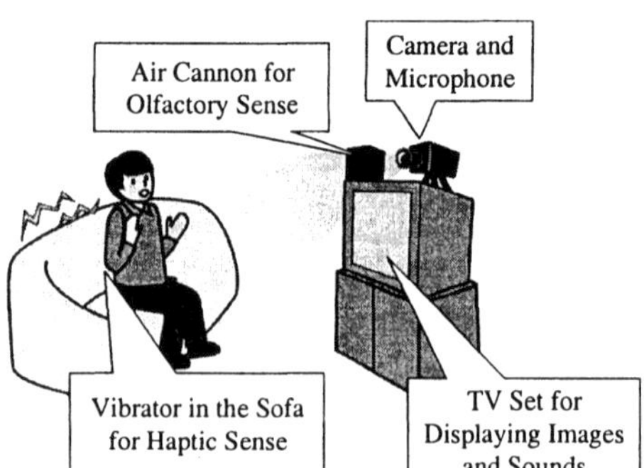

Figure 1. Example of Home Terminal for Networked Interaction Therapy

A system for Networked Interaction Therapy needs to get people with dementia (users) involved and keep their attention. Figure 1 illustrates an example of the home terminal for Networked Interaction Therapy. Currently, we plan to provide our service by using a large-screen TV and a set-top box that controls network communication along with a camera and microphone for monitoring the user's status. We are also considering using an air cannon and a vibrator in the sofa to activate the olfactory [1] and haptic senses [2] in addition to the audio and visual stimuli.

When designing a system for Networked Interaction Therapy and the services provided by the system, it is important to take into consideration the needs of the potential users, in this case people with dementia. However, it is difficult for them to directly participate in the design process. Goodman *et al.* conducted interviews with professional caregivers to assess their patients' needs in a report on designing a health care system for the elderly and disabled [3]. Similarly, we have involved the caregivers, including family and support group members, in our design process. Specifically, we presented our service scenarios in short video clips, which were presented to the caregivers. After the presentation, the caregivers discussed whether our scenarios satisfied their needs and whether they could accept the technology with which we plan to implement the services [4].

Figure 2. Example of 'Coping with Behavioral Problems' Scenario

Our service scenarios cover issues of Internet communication support between people with dementia and support group members, their daily life support, coping with behavioral problems, and so on. Figure 2 shows a scene of the 'coping with behavioral problems' scenario. There are many kinds of behavioral problems affecting people with dementia. Here, we introduced a scenario that addresses wandering at night. This type of behavioral problem is often caused by the anxiety people with dementia feel when they wake up during the night. They may think that they are in a strange place and feel alone because their memories have deteriorated and they do not recognize that they are in their bedroom. In the proposed service scenario, the system detects when they wake up at night and displays an image of a familiar person who talks to them gently to help them cope with problems. In addition, music and aroma may be presented to keep their attention on the system and to offer reassurance.

By using such video clips showing our service scenarios, we studied the need and acceptance of our services. We asked four support groups for people with dementia in Chiba Prefecture, Japan, to watch the video clips and make comments through interviews. According to the results, we found that the caregivers strongly need a system that holds the attention of people with dementia in order to prevent their behavior problems and that they think an audio-visual stimulus is the most effective way to do this [4]. Then, the next question for designing the system is what kind of audio-visual stimulus should be presented to people with dementia in order to hold their attention for enough time to allow the caregivers to rest.

3. REMINISCENCE VIDEO

There has been much research on presenting audio-visual stimuli to people with dementia (users). For example, the concept of 'video respite' was proposed [5], in which a video was created to capture and maintain the attention of people with dementia while giving caregivers time to rest. The respite video often contains a character who talks to the viewer (person with dementia) in a way that attracts his or her attention. Furthermore, Project CIRCA utilized multimedia technology to stimulate the long-term memory of people with dementia [6]. People with dementia have problems with short-term memory even while they often retain long-term memory. This project, in a sense, augments reminiscence therapy with multi-media technologies.

From the viewpoint of stimulating the long-term memory of people with dementia, we selected the so-called 'reminiscence video' as an audio-visual stimulus. This is a slide show video produced from the user's old photo albums. Its effectiveness was experimentally shown [7]. In addition, it is relatively easier for caregivers to use the video once the video is created.

Making a reminiscence video, however, is not an easy task. It involves many steps: first, the photo albums are collected from the user's family. Next, photos are selected from the photo albums. The photos to be used should invoke distant memories of the user. Then, the photos are shot with a video camera, possibly using panning and zooming effects (so-called 'Ken Burns effect' [8]). Finally, a narration is added to get the user more involved in the video. These processes are usually performed by skillful volunteers. They are not the kinds of tasks the family caregivers can do while they look after people with dementia. Moreover, for some users, showing the same reminiscence video again and again may wear off the original effects of the video. Thus, a method for easily making different reminiscence videos from the same set of photos is needed. In order to solve this problem, we have developed an authoring tool that makes use of photo annotations [9]. The annotation attached to a photo includes the theme of the photo, such as 'travel' or 'school', and the date the photo was taken. In addition, 'region' in the photo can be specified by annotations. Each region represents an area of the person's face or interesting objects in the photo. The regions are later used when panning and zooming effects are added. These features are popular for making an attractive slide show video from a set of still photos.

The proposed authoring tool contains a database that stores the scanned photos taken from the user's album. The annotations to the photos are made with the authoring tool. Once we have the annotated photos in the database, we can produce a video as follows. First, we search for the photos to be used in the video by using the annotation data attached to each photo as search keys. Then, the panning and zooming effects are added semi-automatically by using the region data in the photo annotation. Finally, background music can be specified to form a play list.

It is rather time-consuming to add annotations to each photo. However, once we have annotations, we can create various videos from an existing set of photos. This is a very useful capability, especially when the user gets bored easily after viewing the video only once. Photo annotations are mainly used for efficient image retrieval from the image database [10]. In this system, however, annotations are used not only for image retrieval but also for adding suitable visual effects to a reminiscence video.

4. PRODUCING REMINISCENCE VIDEO

Here, we describe how the caregivers (creators) produce the reminiscence video, beginning with attaching annotations to photos. We used the following vocabularies to describe annotations: Dublin Core [11], FOAF [12], and Image Regions [13]. Dublin Core is used to describe various properties of the photo itself. FOAF is mainly used to describe a person included in the photo. Image Regions are used for storing the region data of each photograph. In Figure 3, three regions are defined by using the pointing device, and FOAF attributes for each region are specified via the text dialog.

Dublin Core attributes of each photo can also be added via the text box. The user (video creator) searches for the photos to be used in the video by specifying the keywords that describe the 'theme' of the reminiscence video being produced.

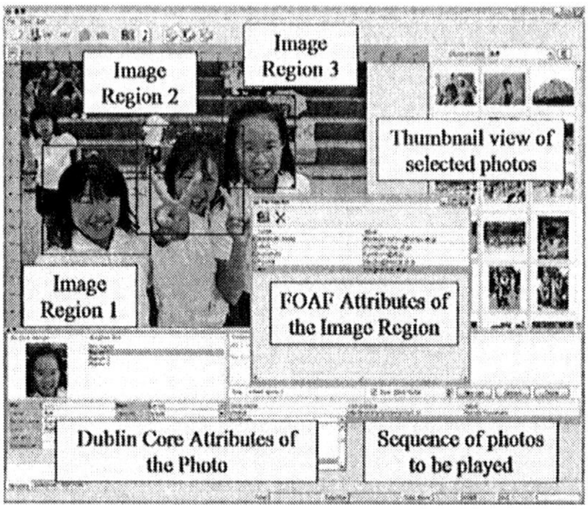

Figure 3. GUI Example of Annotating the Photo

The query can contain regular expressions supported in RDQL [14]. Selected photos are displayed as thumbnail images at the right side of the GUI panel. The creators can drag and drop these images to the list box, where photos for the reminiscence video are listed in the order to be played. There is a template for specifying the kinds of effects and transitions between the photos. We implemented a simple template that uses a zooming effect to a particular region, panning effects between regions, and fade-in, fade-out transitions. In addition, background music (BGM) can be added as a play list. We defined SSVML (Slide Show Video Markup Language), which is a compact representation of a slide show video. With SSVML, we can specify the photos to be used in the video, effects and transitions to be used, and the order of photos to be presented. We also implemented player software that directly interprets the SSVML format.

5. EXPERIMENTS AND RESULTS

We conducted preliminary experiment at users' homes in order to evaluate the effectiveness of the reminiscence video generated by the proposed system. Three male individuals with dementia took part in the experiment. Table 1 shows the users' brief profiles. We prepared four kinds of video clips for each user. The first one is a reminiscence video that is semi-automatically generated from the old photos provided by the family members, and the other videos relate to their hobbies and news clips. Each video clip is about seven minutes long, and we showed these video clips to the users in this order. The BGM (old Japanese nursery songs) was played during each reminiscence video. In order to evaluate the effects of narration, we added narrations only to the first half of each reminiscence video. We used two cameras to monitor the user in this experiment. One is used for recording the user's entire body movement to monitor his overall behavior. The other is for taking the face image of the user to monitor the changes in his facial expression.

	Age	Gender	Patient's History	Problems	Patient's Preference
User A	62	Male	Cerebral Contusion by Accident	*Memory impaired *Verbal Aphasia *Bad-tempered	*Igo (Japanese table game) *Songs of Misora Hibari (Famous Japanese singer)
User B	69	Male	Multiple Cerebral Infarct	*Memory impaired *Bad-tempered	*Watching Baseball Games *Japanese Nursery Songs
User C	81	Male	Alzheimer's Dementia	*Memory impaired *Bad-tempered	*Traveling by Train *Japanese Nursery Songs

Table 1. Users' Brief Profiles

By using these recorded images of the users, we evaluated how much the user was concentrating on the video clip displayed and how he was amused by the video clip. In order to analyze these recorded images, we asked five people, who were not related to the users, to judge each user's levels of concentration and amusement. A subjective scale of one to five was used for scoring. Each subject watched the first minute of the recorded images of the users as well as the last minute and then scored the degree of the user's concentration and amusement for each recorded image.

Figure 4 and Figure 5 show the average scores of users' concentration and amusement for each recorded image of the first and last one-minute segments. As shown in both graphs, the reminiscence video made users more concentrated and amused than any other type of video clips for the first one minute. On the other hand, during the last one minute there seems to be no outstanding difference between the reminiscence video and other types of video materials. One possible reason for this result involves the effects of the narrations attached to the reminiscence video. Users often smiled, nodded, and replied in response to the narrations attached to the reminiscence videos. As mentioned above, no narration was added to the latter half of the reminiscence videos in this experiment. Thus, there is no narration in the last one minute, which resulted in a smaller difference between the reminiscence video and other types of video. It seems that the narration can get the users more involved in the reminiscence video.

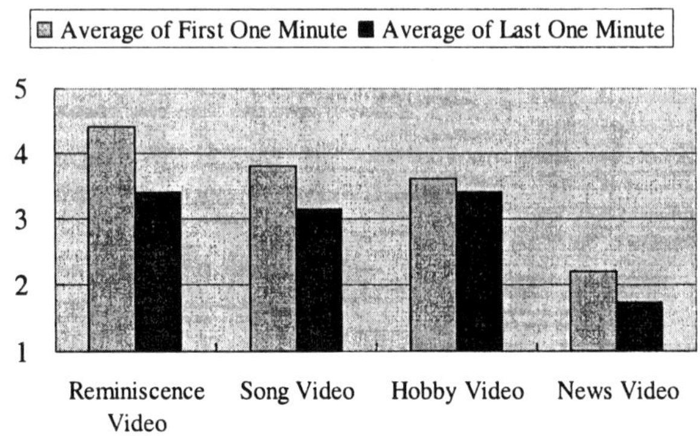

Figure 4. Results of Experiment (Concentration)

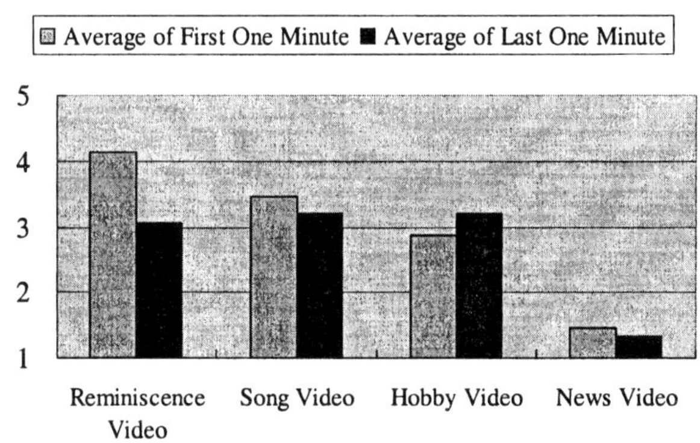

Figure 5. Results of Experiment (Amusement)

6. CONCLUSIONS AND FUTURE WORKS

We introduced a reminiscence video authoring tool that makes use of photo annotations. We also showed how a creator who may not have know-how of video editing could select photos based on the theme of the video to be produced and then create the reminiscence video using a pre-defined template. In the current implementation, we defined only a very simple template. We plan to add various kinds of templates. Also in the current implementation, a creator is required to add each annotation (such as regions in a photo) manually. There have been several research efforts on (semi-) automatic annotation of images (for example [15]). We plan to utilize existing techniques such as face detection algorithms to help a creator specify regions.

In addition, we conducted preliminary experiments to evaluate the effectiveness of a reminiscence video produced by the proposed authoring tool for people with dementia (users). The results show that the reminiscence video made users more concentrated and amused than other types of video clips used in our experiments. The results also indicate that the narration attached to the video may be important. Though further studies are needed, the reminiscence video seems to be a promising stimulus for holding users' attention and for preventing users' behavior problems. We are now conducting a further experiment to evaluate the long-term effects of the reminiscence video. In this latest experiment, caregivers monitor and record the users' behaviors on a regular basis as the users watch the reminiscence video. In this experiment, we will also evaluate how narrations, BGM, and visual manipulations such as the Ken Burns effect influence the ability to hold users' attention and the ability to prevent users' behavior problems. We hope that the reminiscence video will help family members take care of their loved ones and that the proposed authoring tool will make it easier to create the video.

ACKNOWLEDGEMENTS

We would like to thank the participants in the reminiscence video experiments. We would also like to thank Koji Saito for the design and implementation of the authoring tool. This research was supported by the National Institute of Information and Communications Technology.

REFERENCES

[1] Yanagida, Y., Kawato, S., Noma, H., Tetsutani, N., and Tomono, A.: A Nose-Tracked Personal Olfactory Display, SIGGRAPH 2003 Sketches & Applications (2003).

[2] Kakita, M., Yanagida, Y., and Tetsutani, N.: Presentation of A Character Using Distributed Vibro-Tactile Display, IEICE Society Conference A-16-23 (2003) 318 (in Japanese).

[3] Goodman, C.A., Jimison, H.B., and Pavel, M.: Participatory design for home care technology, Proceedings of EMBS/BMES Conference 2002, Vol. 3 (2002) 23-26.

[4] Kuwahara, N., Kuwabara, K., Utsumi, A., Yasuda, K., and Tetsutani, N.: Networked Interaction Therapy: Relieving Stress in Memory-Impaired People and Their Family Members", Proceedings of 26th Annual International Conference IEEE EMBS (2004)
[5] Lund, D.A., Hill, R.D., Caserta, M.S., and Wright, S.D.: Video Respite: an innovative resource for family, professional caregivers, and persons with dementia, The Gerontologist, Vol. 35, Issue 5 (1995) 683-687.
[6] Gowans, G., Campbell, J., Alm, N., Dye, R., Astell, A., and Ellis, M.: Designing a multimedia conversation aid for reminiscence therapy in dementia care environments, Extended abstracts of the 2004 conference on Human factors and computing systems (2004) 825 - 836.
[7] Yasuda, K. et al.: Reminiscence Video for Higher Brain Dysfunctions, Proceedings of General Conference of Japan Society for Higher Brain Dysfunctions, (2004)
[8] http://en.wikipedia.org/wiki/Ken_Burns
[9] Kuwahara, N., Kuwabara, K., Tetsutani, N., and Yasuda, K.: Using Photo Annotations to Produce a Reminiscence Video for Dementia Patients, 3rd International Semantic Web Conference (ISWC2004) Demo Papers (2004).
[10] Schreiber, A.T.: Dubbeldam, B., Wielemaker, J., and Wielinga, B.: Ontology-based photo annotation, IEEE Intelligent Systems, Vol. 16, No. 3, (2001) 66-74.
[11] http://dublincore.org/
[12] http://www.foaf-project.org/
[13] http://www.bnowack.de/w3photo/pages/image_vocabs
[14] http://www.w3.org/Submission/RDQL/
[15] Suh, B. and Bederson, B.B.: Semi-Automatic Image Annotation Using Event and Torso Identification, Tech Report HCIL-2004-15, Computer Science Department, University of Maryland, College Park, MD (2004)

AUGMENTING PHOTOGRAPHS WITH SOUND FOR COLLOCATED SHARING

An exploratory study of the Audiophoto Desk

Siân E. Lindley and Andrew F. Monk
University of York, U.K.

Abstract: The Audiophoto Desk is a tangible interface. It plays a sound that has been associated with a photograph when a printed version of that photograph is placed on the desk's surface. This study explored how three groups of four friends used the desk to share photographs in the context of two types of photo-talk: reminiscing and storytelling. The study had three aims. The first was to investigate the types of sound that users would choose to accompany their photographs. In particular it was of interest to see whether recipient design would be relevant in explaining why the participants chose some of their sounds. The second aim was to examine the effect that different types of audio would have on the conversations during photograph sharing. Music, sound effects and voiceovers were found to have different effects on the latency before the conversation began and the extent of the conversation. The third aim was to understand the user experience of the groups. Despite the use of printed photographs, some participants felt that the Audiophoto Desk made photograph sharing too formal a process for the home.

Key words: Photographs, music,

1. INTRODUCTION

A recent development in digital photography is the introduction of cameras that can record sound as well as images, producing 'audiophotographs' (Frohlich & Tallyn, 1999). The sharing of this type of photograph necessitates some form of technology to allow the sounds to be played back, something that is possible using photoware for the PC. However, a number of researchers have noted that while the advent of digital photography has made it easier to share photographs remotely, it is less appropriate for supporting face-to-face conversations, largely due to the limitations of available display technologies (e.g., Balabanović *et al.*, 2000; Shen *et al.*, 2002). This may be one reason why users tend to print out their digital photographs for the purposes of sharing them (Frohlich *et al.*, 2002, Crabtree *et al.*, in 2004). This study explores how groups of friends used an interface that has been designed to support the sharing of 'audioprints' (Frohlich *et al.*, 2000): the Hewlett Packard Audiophoto Desk (Frohlich *et al.*, 2004). This is a tangible interface that plays sounds to accompany printed photographs, and enables users to share audiophotographs without sitting around a computer monitor. Photographs that are placed on the desk's surface are recognised by a computer via a webcam. The computer then plays the appropriate sound to accompany the photograph. However, from the user's perspective, the computer plays a minimal role in the experience. This study aims to explore a number of aspects of audiophotography in the context of photograph sharing using the Audiophoto Desk. Whereas previous studies have used audio CDs, PC photograph albums and even handheld audioscanners (Frohlich *et al.*, 2000) to enable groups to play back sounds, no other research known to the authors has used the Audiophoto Desk in this way.

Previous work on audiophotography has used various types of audio, which have been used for different purposes. Research by Frohlich and Tallyn (1999) shows that ambient sounds recorded at image capture were perceived as enlivening photographs and acting as triggers for memory. Further work by Frohlich (2004) shows that simply recording conversations about photographs can be useful, making the photographs themselves more personalised and providing details that might otherwise be forgotten. Here, audio can be seen as an alternative to the written annotations often found in photograph albums, but one that is easier to create than explicit verbal annotations. However, it is also noted that:

> "Participants ... were most reticent of recording voiceover when they could not imagine an appropriate audience. In fact the most difficult audience to record for was oneself, since participants already knew the story of their photographs and found it hard to anticipate what details they might forget." (page 165)

A study of a portable device that allows audio to be recorded alongside sequences of photographs (Balabanović *et al.*, 2000) confirms this finding. Users did not record vocal annotations for their own purposes, instead envisaging voiceovers as a way of supplementing photographs to be sent to others. Moreover, users did not take the opportunity to record their own photo-talk, although they did note that new behaviours could evolve over time.

The main aim of the present study is to explore the types of sound that will be chosen by users to accompany their photographs. Here, audio will be chosen to accompany existing photographs rather than recorded at image capture, and the users will know who they will be sharing their photographs with when they select their sounds. This may help overcome some of the problems in choosing sounds noted by Frohlich (2004), in that users will be able to design their sounds for a particular audience. The notion of designing audio for known listeners can be seen as an extension of the recipient design found in ordinary conversation (Sacks *et al.*, 1974) and also in photo-talk (Frohlich *et al.*, 2002). Frohlich *et al.* note that conversations about photographs often represent a trade-off between the stories that the photograph owner(s) want to tell and the interests of their audience. It is hypothesised that the selection of audio in this study will reflect the interests of the audience that it has been chosen for.

A second aim of the study is to investigate what effect various sounds have on groups' conversations. It is hypothesised that ambient sounds, such as background music or the sound of the sea, will encourage conversation, whereas audio that requires attention, such as verbal annotations, will inhibit it. A third aim is to investigate how the Audiophoto Desk is perceived and, in particular, whether it is seen as an appropriate domestic technology. The groups' perceptions of how compatible this type of technology is with the informal approach to photograph sharing that is adopted in the home will be considered in this.

The study involved three groups of four friends. The groups were split into pairs, with each pair being asked to provide a set of photographs to be shared with the rest of the group. One pair was asked to bring in a set of photographs of an event that involved the whole group, to allow reminiscing to occur (see Frohlich *et al.*, 2002). The other pair was asked to bring in photographs of an event that one of them had done without the others, allowing storytelling to occur. The pairs were also asked to choose a sound to accompany each photograph. They were provided with a sound effects CD and a CD of ambient music to help them to do this, but they were also able to choose music from their own collections. Additionally, they were supplied with a MiniDisc recorder so that they could create their own sound effects and voiceovers. Their behaviour while sharing the photographs, and their perceptions of that experience as measured by questionnaires and interviews, form the basis of the study.

2. METHOD

2.1 Participants

The participants were three groups of four friends, who were all undergraduate students at the University of York. There was one all male group, one all female group, and one mixed group of one male and three females. Each group had known each other for at least a year prior to the study. The mean age was 20.70 years, with a standard deviation of 0.71 years.

2.2 Materials

The study was run in the dining room of the York Responsive Home, a smart home recently developed at the University of York. The Audiophoto Desk was set up on the dining table. The groups' behaviour was recorded using an unobtrusive camera on the opposite wall. The setup of the room is shown in Figure 1.

Figure 1. Experimental setup

Details of the desk's functionality are as follows. The Audiophoto Desk plays a sound that has been chosen to accompany a photograph when that photograph is placed on the desk's surface. The sound continues to play until the clip ends or until the photograph is removed. If two or more photographs are placed together, the accompanying sounds will be played simultaneously. Additionally, volume can be altered by moving photographs towards the front of the desk (increasing volume) or towards the rear (decreasing it), and sound can be panned across the two speakers on the surface of the table by moving photographs from left to right. When photographs are obscured from the camera and then revealed again, the associated sounds restart.

Two questionnaires were administered to the participants during the study. The first gathered background information about the participants and their friendships with the other participants in their group. The second examined what the participants thought of the Audiophoto Desk by asking them to rate their agreement with a number of statements using a five point Likert scale. The questionnaire addressed five topics. There were two statements for each topic, one positive and one negative. The five topics were as follows:

- Ease of use
- Satisfaction with experience
- Perception of experience as fun
- Involvement within the group
- Perceived enhancement of photograph sharing.

2.3 Procedure

The study was split into three sessions for each group.

Session 1. At the first session the Audiophoto Desk was demonstrated to the group. The group was asked to form two pairs, and each pair was instructed to bring at least ten, but no more than 18, of their own photographs with accompanying sounds to the next session. This was because the Audiophoto Desk is limited to storing the information for 18 photographs and their sounds at any one time.

Session 2. The second session was run with each pair separately. This required the participants to bring in the photographs that they had chosen and the accompanying sounds, as well as a list to clarify which sound was to accompany each image.

Session 3. In the final session the participants were seated at the Audiophoto Desk in their two pairs. They were reminded how to use the Audiophoto Desk and then asked to share with their friends the photographs that they had chosen. Following this, each group member completed the questionnaires and was interviewed individually.

3. RESULTS AND DISCUSSION

Videos of the first eight minutes of the groups' behaviour for each set of photographs were examined. The value of eight minutes was chosen because the group who had the shortest conversation talked for just over eight minutes. Events such as conversational turns, placement of photographs on the desk and audio playback were coded using 'The Observer', a software package that supports video analysis.

The results will be presented in accordance with the aims of the study. Data on the types of sounds that were chosen and their effect on the conversation will be given first, followed by questionnaire and interview findings to allow an insight into the users' experiences.

3.1 Types of Sound used to Augment Photographs

The sounds chosen by the participants to augment their photographs could be placed into three categories: music, sound effects and voiceovers (voiceovers were classified as containing spoken words). Sound effects and music could either be created especially by the participants or chosen from existing CDs. The types of sound chosen by the participants to accompany their photographs are given in Table 1, and some examples of each category are given in Figure 2

Table 1. Types of sound used to augment the 66 photographs totalled over the 3 groups, classified as created especially for the photograph or taken from existing recordings.

	Music	Sound effect	Voiceover
Created	2	6	17
Existing recording	23	18	-

Table 1 shows that whereas voiceovers were (by necessity) created especially for the photographs, music and sound effects were more likely to be taken from existing recordings. This is unsurprising, given that music is harder to create, and that the participants had their entire CD collections to choose from. The distinction between created sound effects and voiceovers was sometimes quite tenuous, with sound effects such as wolf-whistles being put in the sound effect category simply because they involved no words. An alternative approach to creating a sound effect is described below.

The photographs in Figure 1 illustrate some of the different ways in which sounds were used to augment the images. In some cases audio was used simply to enhance the scene shown in the photograph. For example, the sound of someone making a drink seemed to be created to represent the type of background noise that would have been present during a night out. Atmosphere could also be created using music. Chicane's "Offshore" was chosen as an upbeat piece of music that reflected and emphasised the action captured in the image. In these cases audio was used in a similar way to that found by Frohlich and Tallyn (1999), in that it enlivened the scene shown and helped trigger memories. For one photograph, accompanied by "Happy Birthday", the music acts as a clear reminder of the depicted event.

	*Voiceover: Created* Reading of Dante's "Midway along the journey of my life"
	*Sound effect: Pre-recorded* Toy plane
	*Music: Pre-recorded* "Offshore" by Chicane
	*Voiceover: Created* Reading of Dante's "Midway along the journey of my life"
	*Sound effect: Pre-recorded* Toy plane
	*Music: Pre-recorded* "Offshore" by Chicane

Figure 2. Examples of photographs and their accompanying sounds for each category.

However, audio that had a more tenuous connection to the photograph was also sometimes chosen. Such sounds triggered memories that were not directly relevant to the events shown in the photograph. For example, the sound of a toy plane was chosen because one of the people in this photograph liked planes and had been known to the photograph owner since childhood. In this way sounds could be very personal and largely meaningless to people who did not know the background to them (including other members of the group). The explanation of why these sounds had been chosen supplemented the conversation, and introduced topics that might not otherwise be covered in typical photo-talk. However, the majority of photographs were augmented with audio that was at least meaningful to the people within the group. Many of these seemed to be 'in-jokes'. For example, the use of the word "Legend!" to accompany one photograph was met with a great deal of laughter, and seemed to be a catchphrase of the person shown. In cases where the audio did not represent a known joke to the group, it was still often chosen with the other group members in mind. For example, the Dante reading in Figure 2 was chosen to reflect the "calm before the storm" depicted in this photograph, and once this was explained it seemed perfectly suited to the group's sense of humour. Similarly, much of the music that was chosen was selected because it suited the tastes of the group that would be listening to it.

One type of audio that was conspicuous in its absence was the type of voiceover that involved someone explicitly describing the scene. There were only two instances of this (both from the same pair), both of which were greeted with laughter and were seemingly inappropriate for the task of sharing audiophotographs as a collocated group. It can be seen that in addition to the participants showing recipient design in their choice of sounds, they also largely chose audio that would be suited to the situation of photograph sharing, and avoided vocal annotations that might be more appropriate in a photograph album.

3.2 Effect of Sounds on the Conversation

To give an indication of how long and how often different types of sound were listened to, the duration for which they were played and the number of instances of a sound being restarted are given in Table 2. As these statistics are totalled over the photographs viewed by all three groups in the first eight minutes of conversation, not all of the audiophotographs that were developed are included. The number in each category used in the analysis is given in first row of the table.

Unsurprisingly, music was played for longer than any other type of sound before either reaching its end or being removed from the desk. The music clips tended to be the longest and therefore obviously had the potential to play for longer. Voiceovers tended to be very short, but there were a couple of very long voiceovers, for example poetry readings, that have caused the mean to be higher than is representative of the data. The median gives a much more accurate indication of voiceover duration.

It was quite rare for any sound to be deliberately restarted. This may be partially due to the fact that once audio clips had stopped they would restart spontaneously if they were briefly obscured from the camera while the groups were using the desk. Voiceovers seemed to spontaneously restart much more often than any other sound, probably because, due to their short duration, they would be left at the front of the desk, where the potential for being overshadowed was greater. Participants tended to restart sounds for the fun of mixing them together, rather than for the sake of listening to them again. When mixing audio, they appeared to be fairly indiscriminate about what they were mixing, simply grabbing photographs that were within reach.

Table 2. Mean (with standard deviations) and median duration that different types of sound were played for, and the total number of instances that photographs from each category were restarted

	Music (N = 20)	Sound effect (N = 16)	Voiceover (N = 10)
Mean duration of sound (s)	67.27 (57.75)	22.37 (20.28)	24.99 (38.89)
Median duration of sound (s)	45.9	13.7	3.24
Total that were deliberately restarted	3	6	1
Total that spontaneously restarted	6	5	15

Table 3. Mean scores (and standard deviations) for the total amount of time the group spent making conversational turns and the latency before the conversation begins, when talking about photographs accompanied by different categories of sound

	Music (N = 19)	Sound effect (N = 11)	Voiceover (N = 10)
Mean latency before the conversation begins (s)	2.64 (2.66)	3.04 (1.78)	6.82 (9.43)
Mean total amount of time spent in turns (s)	42.41 (31.29)	32.62 (17.82)	37.13 (21.65)

To try to understand the overall effect of the different categories of sound on photograph sharing, the length of the conversation (measured by summing the total amount of turns in seconds, but not backchannels, laughter or side comments) and the time spent listening to the sound before any turns are taken are given in Table 3. Because only photographs that had been fully discussed within the eight minutes were used in this analysis, the number in each category is slightly lower than in Table 2.

As Table 3 shows, latency before the conversation begins is much shorter for music and sound effects than for voiceovers, as predicted. Voiceovers must be attended to if they are to be understood, resulting in the conversation being suspended while they are playing. As shown in Table 2, the median duration of the voiceovers is 3.24 s, so it seems that in many cases the participants listened to the full sound before the conversation began. In contrast, music tracks, which are most likely to be familiar to the group, are quickly recognised, allowing the conversation to begin.

Music is associated with more time spent in conversational turns than the other two sound categories. The fact that music clips were listened to for the longest amount of time may have encouraged the conversation to go on for longer by making silences less awkward. Furthermore, the end of a clip often signalled a chance to introduce a new sound and therefore move the conversation on. The fact that this was delayed with music clips may also have contributed to the longer conversations. It was hypothesised that sound effects would be similar to music in encouraging longer periods of conversation compared to voiceovers. In fact, voiceovers stimulated more time spent in conversational turns than sound effects did. Voiceovers were more likely to be meaningful, as opposed to simply providing background noise, than sound effects, and it may be the case that this stimulated more conversation. However, it should be noted that the variance across the different categories is extremely large, making it possible that the pattern found is not reliable.

3.3 Interview Findings and Observations relating to Different Types of Sound

Participants were generally positive about the addition of sounds to photographs, saying that it "made it more enjoyable" and "added humour". However, some of the sounds were deemed to be more suitable than others by the participants. Music was popular overall and could easily be integrated into the conversation. One participant said, "Using music's better because the voices distract you more". Another disadvantage of voiceovers compared with music was that they tended to be very short. One participant commented, "It would have been cool if we'd had more music files going on in the background and we could have talked over them, instead of waiting for this three second sound but then it would be gone".

An unexpected finding was that listening to the audio sometimes took precedence over looking at the photographs: "Some photos were probably preferred more than others, because of the [sounds], you know they were funny or the song was good, so maybe the other ones got discarded".

On some occasions the addition of sounds was thought to detract from the experience of photograph sharing. Sometimes this was related to the functionality of the desk, which had a slight delay before a sound played. One participant said, "Sometimes you'd put the photograph down and then you were waiting for the sound effect to come up, or things got confused and lost, and you know something didn't come out so [you] wanted to put it down again". This seemed to be an extra burden on the people who had provided the photographs, who seemed to feel in some way responsible for the smooth running of the experience. One of these participants commented, "We were very conscious of the way the pauses kicked in".

The use of sound also made it difficult for participants to hold photographs to get a closer look once they had been placed on the desk; "If you picked them up you interrupted the sound effect and if you put them down again they started all over again". Some of the sounds that were recorded were thought (in hindsight) to be inappropriate for the actual sharing of photographs. This was particularly true for voiceovers giving information about the photograph. As one participant said, "Like saying 'this is me with so-and-so', that didn't really work cos he was here so he could say, 'this is me with so-and so'".

Participants felt that adding sounds "made the conversation a lot easier" and made the experience of photograph sharing "more personal". One participant said that it "made you engage a bit more because immediately you start to try to think of the link which means that you look at the photo", and another commented, "Because the sounds were from different people it kind of brought them into the experience of it more". However, not all of the participants thought this. One said, "I wouldn't say it makes you read anything more into the photo or really look at the photo in a different way".

3.4 User Experience

3.4.1 Questionnaire Results

The results of the questionnaire are shown in Figure 3. The scores for the positive and negative statements on the same topic were combined by subtracting the score for the negative statement from that for the positive statement. This resulted in all scores falling on a scale from –4 to 4, where a positive score represents a positive experience.

Figure 3 shows that all groups rated their experience positively, with no scores falling below 1 on the scale of –4 to 4. In general, the highest ratings were given to questions about the experience being fun and relatively low ratings were given to questions about satisfaction. It is possible that the lower ratings for satisfaction may be related to the participants' expectations of the desk. As the initial demonstration of the desk used photographs that were designed to be played together, and was performed by an experienced user, the technology appeared to be quite seamless. This was not always the case when the participants themselves used the desk for photograph sharing.

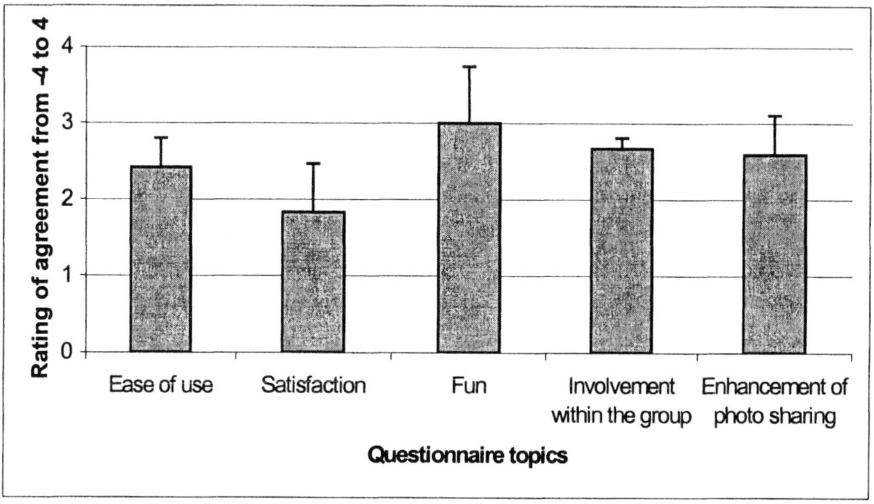

Figure 3. Graph to show the mean questionnaire results. The error bars indicate standard deviations (N = 12).

3.4.2 Interview Findings and Observations relating to User Experience

Size of display. Having one photograph at the front of the desk meant that everyone was able to look at the same photograph at the same time. Most of the participants felt that this was a positive aspect of using the desk, with one saying, "Normally when you share photos you look at it and then you pass them to the next person, so you're only describing it to the next person, rather than everybody being able to look at it". However, not everyone accepted the positive aspects of viewing one photograph centrally. A contrasting view was that "it's a bit strange the way we had four of us sitting

round [...] when you hold it in your hands you've got people in closer, it's a closer experience". This may be another reason why the ratings for satisfaction were relatively low; although the desk was perceived as fun it may not have been seen as supporting the intimacy normally found in photograph sharing.
Interacting with the desk. A number of the participants developed misconceptions about how the desk worked. In particular, restarting sounds proved to be difficult. One said, "we were placing it down and kind of pressing it down quite firmly so that it would play", and another noted that "when you tap them they come back on again". In fact neither of these strategies were correct; to restart a sound it was necessary to hide the photograph from the camera before revealing it again. Members of the same group tended to share similar misconceptions about how the desk worked, and this may account for the high ease of use ratings given in the questionnaire; clearly they felt that they understood how to use the desk. The difficulties that the participants had seem to be a problem of this tangible interface not actually being responsive to touch (Lindley & Monk, 2004).

Using the desk in the home. The participants were asked how they would feel about using the desk in their own home. The general feeling was summarised by one participant who said, "It would have to be a fairly special thing for you to set it all up, but if you were doing a presentation, or maybe showing some people you didn't know [...] then I think it would be good". Most of the participants felt that using the desk added a formality to photograph sharing; "It sort of comes across as a bit like a presentation in the work place". One said that it would be "a bit forced, to say to people let's come round and look at my photos". Another issue was the amount of time viewed as necessary to set up the Audiophoto desk. It was viewed as "quite labour-intensive, a bit time-consuming" and therefore something that might be used "now and then, but not regularly". One participant summarised the feeling of audio sometimes being unnecessary in photography by saying, "It would have to be a very special occasion... I wouldn't take out a video camera for instance and video a night out with the guys".

4. SUMMARY AND CONCLUSIONS

The main aims of this study were to investigate the audio that users would choose to accompany their photographs and to explore the effect of different sounds on the conversation. It was hypothesised that recipient design would be reflected in the selection of audio and that whereas ambient sounds such as music and sound effects would encourage talk, voiceovers would inhibit it. An additional aim was to understand how the Audiophoto Desk would be perceived by the participants who used it; particularly in terms of its appropriateness as a domestic technology.

The results generally support the first hypothesis. While some sounds were selected to serve the purpose (previously noted by Frohlich and Tallyn, 1999) of enlivening images and triggering memories, most audio was chosen specifically to suit the audience. Music was often selected with the tastes of the group in mind, and voiceovers capitalised on in-jokes and knowledge that was common to the group. Explanations of why these sounds were amusing were left unspecified. This is analogous to the way that common memories associated with photographs are often left unelaborated (Frohlich *et al.*, 2002), and can be seen as a way of demonstrating common ground (Clark & Brennan, 1991) and friendship. In addition to designing sounds for a particular audience, the fact that photographs would be shared with a collocated group also seemed to have an impact on the participants' choice of sounds. The rare occasion of photographs being augmented by explicit vocal annotations was generally seen as being inappropriate, because it did not make sense in the context of collocated photograph sharing. The fact that only two out of 17 voiceovers consisted of this type of content indicates that audio was generally chosen to suit the situation.

In considering the second hypothesis it does seem that the different categories of audio have different effects on conversation. Although the small number of groups prevents statistical significance from being verified, the intent of the study was simply to explore possible effects of sound on behaviour. The results were surprising in that while music encouraged the longest conversations, the shortest conversations occurred when accompanied by another type of ambient sound: sound effects. It seems that although voiceovers did delay the beginning of the conversation, they were short enough for this to not be detrimental. Furthermore, it may be the case that their meaningful content actually promoted a good deal of talk.

The third aim of the study was to consider how appropriate the Audiophoto Desk is as a domestic technology. Surprisingly, the Audiophoto Desk was not seen as offering the flexibility of printed photographs, despite its use of audioprints. Instead the large display was perceived as too formal for the home. It may be the case that portable audioscanners, such as the one used by Frohlich *et al.* (2000), will be more able to maintain the advantages of printed photographs, because they can be easily incorporated into ongoing discussions or activities. Additionally, the task of selecting sounds for each image was seen as too much effort. This is something that may be overcome with the use of digital cameras that record sound, but obviously needs to be addressed if sounds are to be tailored for audiences and situations as they were in this study.

The findings of this study have a number of implications for audiophotography. It should be noted that while vocal annotations are inappropriate for photograph sharing, and sound effects that add atmosphere are the least likely to facilitate conversation, these are the two types of sound most likely to be recorded with digital cameras. While these sounds may enhance the experience of looking at photographs in an album and be an enjoyable supplement in remote sharing, they are less suited to photograph sharing by collocated groups. This study has shown audiophotography's potential for recipient design and its capacity to provide a fun experience when photographs are shared by groups. However, to capitalise on this potential the development of technology that makes it easier to augment photographs with more than one sound, and to share them in a way that maintains the informality of printed photographs is necessary.

ACKNOWLEDGEMENTS

We would like to thank John Robinson, at the University of York's Department of Electronics, for setting up the Audiophoto Desk. We would also like to thank David Frohlich, at Bristol's HP Labs, for useful discussion regarding the design of the study.

REFERENCES

Balabanović, M., Chu, L.L. & Wolff, G.J. (2000). Storytelling with Digital Photographs. *CHI Letters, 2 (1)*, 564-571.

Clark, H.H. & Brennan, S.E. (1991). Grounding in Communication. In L.B. Resnick, J. Levine, & S.D. Teasley (Eds.), *Perspectives on Socially Shared Cognition*, 127-149. Washington DC: American Psychological Association.

Crabtree, A., Rodden, T. & Mariani, J. (2004). Collaborating around Collections: Informing the continued development of photoware. *Proceedings of CSCW '04*, 396-405. Chicago: ACM Press.

Frohlich, D.M. (2004). *Audiophotography: Bringing photos to life with sounds*. London: Kluwer.

Frohlich, D.M., Adams, G. & Tallyn, E. (2000). Augmenting Photographs with Audio. *Personal Technologies, 4 (4)*, 205-208.

Frohlich, D.M., Clancy, T., Robinson, J. & Costanza, E. (2004). The Audiophoto Desk. *Proceedings of 2AD, The Appliance Design Conference*. Bristol: Appliance Design Network.

Frohlich, D.M., Kuchinsky, A., Pering, C., Don, A. & Ariss, S. (2002). Requirements for Photoware. *Proceedings of CSCW '02*, 166-175. New Orleans: ACM Press.

Frohlich, D.M. & Tallyn, E. (1999). Audiophotographs: Practice and prospects. *Extended Abstracts of CHI '99*, 296-297. New York: ACM Press.

Lindley, S.E. & Monk, A.F. (2004). DJing with Photos. *Proceedings of HCI 2004, Volume 2*, 179-180. Leeds: Research Press.

Sacks, H., Jefferson, G. & Shegloff, E. (1974). A simplest systematics for the organization of turn-taking for conversation. *Language, 50 (4)*, 695-735.

Shen, C., Lesh, N.B., Vernier, F., Forlines, C. & Frost, J. (2002). Sharing and Building Group Histories. *Proceedings of CSCW '02,* 324-333. New Orleans: ACM Press.

EXPLORING EXPRESSION OF FORM, ACTION AND INTERACTION

Steven Kyffin, Loe Feijs, and Tom Djajadiningrat
Steven Kyffin, Philips Design, Eindhoven, Netherlands, Loe Feijs, and Tom Djajadiningrat, Eindhoven University of Technology, Netherlands

Abstract: The central theme in Ambient Intelligence is that powerful computation, communication and storage facilities are available, but are invisible. We believe that *objects* will play an important role as mediators. In order to make models of mediation and to study the mediation of behaviour, traditional modelling techniques, such as foam models, rendering etc. fall short. Although it is easy to make two-dimensional models of screens and front-panels and connect actions to the corresponding buttons, this approach is inadequate for new types of interfaces and behaviours which are not screen-based. Therefore we have experimented with an approach which we call 4D Sketching. The term 4D refers to the fact that it includes sketching in 3D space with the active behaviour adding a fourth, temporal dimension. We use foam, cardboard, a glue-gun, and other easy-to-use materials for the spatial sketching. We use a microprocessor, servo-motors and sensors to sketch active behaviour. The article describes the underlying philosophy, the technical aspects and a number of experiences gathered so-far, including student work from a project called Semotion.

Key words: Emotions, sketching, model making, software, product semantics, Movement, Motion, Dynamic behaviour, Robotics.

1. INTRODUCTION

The vision of ambient intelligence as put forward by Weiser and adopted by ISTAG and many companies, forms the basis of considerable R&D efforts. These efforts will affect and shape the future of home information systems and eventually the homes themselves and the ways of being of people at home. Examples such as the GeorgiaTech's aware home, the Philips Home Lab in Eindhoven, the inHaus in Duisburg, the OpenHomeAsia project in Singapore, the Equator project in the UK are carriers to explore ambient intelligence. The central theme is that powerful computation, communication and storage facilities are available, but are invisible. In Marzano's "La Casa Prossima Futura" the black boxes have disappeared and the living room contains objects and furniture, again. Then if the traditional terminals disappear, what are the mediators between people and this hidden intelligence? How do people control, and get feedback from, these resources in a way that is meaningful and even attractive at a human level (not just efficient at a machine level).

We believe that *objects* will play an important role as mediators, (besides automatic detectors and wall-mounted displays). In the simplest cases, ordinary objects get tagged or equipped with buttons and displays. But to take full advantage of the richness of human-object interaction and to use the potential of affective (emotional) interactions, there is a need for a new approach. This is an important aspect of home information systems that deserves more attention, also by industrial designers. At present, there is an emphasis on the input rather than the output side: sensor technology, video recognition and object tracking. Also, in research into affective computing and products there is emphasis on the recognition of emotions rather than affect expression e.g. Picard at MIT and Wensveen at TU/e. Most often the output is restricted to flat displays (or, perhaps in Japan, humanoid or animal-inspired robots). An interesting exception is the key table and picture frame developed in the Equator project. Although many traditional home products such as toasters and coffee cans, which do have mechanically moving parts, follow a trend to converge with computing, this convergence comes in the form of electronic displays and buttons being added to traditional forms. As an alternative we explore adding behavioral expression to the existing movement possibilities.

We envisage the need for new types of processes and tools to support the creation of the envisaged new product types. In our work we focus on product behaviour that is enriched with physical movements. Several possibilities exist: either the product is moving anyhow, or the movements are added just for the sake of communication. In both cases, the designer has considerable freedom how to shape the movements and the interactions.

Forms, either concrete or abstract, always carry meanings. It is the responsibility of the designer to make good use of these meanings, for example, to make products beautiful, to stress the importance of certain values, or to improve a product's ease of use and to promote or negotiate enriched experiences between people (communities) and people, people and objects and in time between objects (systems of objects) and objects. Design uses its own languages for this purpose, just as poets, painters, journalists, sculptures, film makes and so on do. The topic of this paper is how to explore the combined usage of form, colour and behaviour as a design language. We have to look for new ways of exploring behaviour, not separately, but in relation to traditional forms. The 4D sketching approach is helpful to find out in how far object behaviour can be used as a language and in how far it can communicate meaning.

Let us define sketching as the creation of form proposals that are:

- fast
- ambiguous

A sketch is *fast* in order to allow the designer to evaluate the proposal and let others evaluate the proposal while still having time for modifications and new proposals. A sketch is *ambiguous* if it does not suggest a higher level of precision than intended by the designer in a certain early stage of the development process. Ambiguity prevents an evaluator from focusing on details that are not supposed to be fixed yet. In other words: sketching can be used to explore a language of forms and their meanings before making the real decisions.

In this paper we show how sketching can be extended from 2- and 3-dimensional sketching to include behavioural aspects too.

2. TOWARDS RICHER MODELLING

New technologies have led to a redefinition of the concept of 'object'. Those objects, which exist only inside the computer, are referred to as 'virtual', and for the time being it is still impossible to give them a physical presence, because there are not yet any materials that conform to their forms. However, these objects do live within the realms of another physicality, a host, so to speak. It is the seamless integration or convergence of the two levels of existence that design needs to begin to reconcile. One of the `raw materials' the designer will be using in the future consists of software technology.

Whereas the `old' designer would use 3D forms, colours and materials, we envisage a need for designers who can use communication and software since these are the `raw materials' for the design of the virtual. Designers and students of design continue to use a multitude of different media to 'draw' out, illustrate and explore their design ideas. Two dimensional, representational techniques, using paper, pens and inks, have often proved sufficient for representing intended ideas. The introduction of computer-based mediums has accelerated and brought realistic accuracy to the process.

The computer has also given rise to the integration of the two and three-dimensional aspects of the 'formal' aspects.

In the past designers would typically draw ideas on paper and explore form, mass, scale and weight based issues in the third dimension, eventually bringing the two aspects together in a single object as an appearance model. Today the computer enables both issues to be tackled simultaneously with the added advantage that through rapid prototyping techniques, such as stereo-lithography and five-axis milling, a 'hard' representation can be prepared with ease. The geometry data may also be used in the preparation and development of the final production piece without the need to 'rebuild' the form in the computer after the visual and feasibility testing is complete.

However, as designers now face an ever wider spectrum of 'designed' requirements, now including software behaviour, content flow, and even hardware behaviours it is imperative that new forms of drawing or even "sketching' are developed to enable the illustration, discovery and exploration of all these 'conflicting' functional needs in order to ensure the full integration of each effecting the other.

These forms of drawing need be no less 'physical' than those currently used, but probably need to merge both the physical and the 'virtual' elements simultaneously.

Having agreed this premise, the primary issues are: How does one draw this materiality? What are the media of this drawing? How do we combine all the elements without actually making the finished 'thing'? How can we make it fast enough to realise something without actually loosing the idea or inspiration in the realisation process? Will the means be strong enough to encourage others to not only understand the idea but to co-laborate in its 'birth'. Are there means of drawing these new things, which enable others to join in as co-creators?

The examples in Sect 6 will demonstrate that not only have the designers used the form of drawing we call 4D sketching as a means of illustration of a mental model of an idea, but in a very rich way the drawing process has actually released ideas which has enriched and in one case enabled the designer to surpass the mental model altogether.

The three dimensional aspects of the 'drawing' is carried out using 'easy to model' traditional cardboard, foam, paper and small found objects (see Sec. 3.3). The fourth dimension or the animated life drawing is explored using simple electrical, electronic and mechatronic components coupled with the materials from the third dimension (see Sect. 4). Our ambition is to encourage and enable the designers and students of design to draw at all levels from the outset and even to draw as a community rather than as a single person.

3. THE TRADITIONAL DESIGN PROCESS: IN HOW FAR DOES IT SUPPORT EXPLORATION OF EXPRESSIVE MOTIONS?

One of the new challenges in developing the physical and virtual aspects of products simultaneously, is that—as products gain robotic characteristics—autonomous behaviour may no longer be limited to changes in two-dimensional displays, but may result in physical, mechanical, three-dimensional movements. Traditional industrial design education has taught few, if any, methods or techniques for exploring the expressiveness of such movements. Clearly, expressiveness of appearance has always been an important issue in industrial design. Techniques aimed at exploring appearance—including making moodboards, sketching and pre-modelling—have become established parts of the design process (for an overview see Hummels et al. [4]). Expressiveness of movements, however, is a different matter. Unlike dance or theatre, traditional industrial design has had little need for exploring dynamic expressiveness. We are interested in the semantics of movement, analogous to how in the mid-eighties there was an interest in the semantics of appearance (amongst others Krippendorff & Butter [5]; Aldersey-Williams et al. [1]). In this section we present a design process that we put together with a view to exploration of dynamic expressiveness. It combines techniques for exploring expressiveness of movement with techniques for exploring expressiveness of appearance:

- collecting video of autonomous mechanical movements
- moodboards
- sketching
- pre-modelling
- acting out (full body movements)
- acting out (hand movements)
- 4D sketching (explorative)
- 4D sketching (presentational)

Below we will discuss the potential role of each technique, and illustrate some techniques by work from two projects. The first project, by Feijs and Kyffin explores the potential of a telephone cradle for mediating the urgency and valence of incoming phone calls. The second project was to raise our awareness of in how far existing design techniques support exploration of dynamic expressiveness. We set up a second year bachelors project at our faculty in which we guided five students in a design process with behavioural expressiveness in mind. In twelve weeks, each student had to design a machine that was to offer its user a walnut, expressing a particular emotion in both its appearance and mechanical movement. The emotions were love, anger, enthusiasm, shyness and panic. In the remainder of this section, we provide an overview of some of the more traditional design techniques and our understanding of how these techniques support exploring expressiveness of movement and expressiveness of appearance. The actual 4D sketching technique is discussed in section 4.

3.1 Video of Mechanical Movements

The student project started with collecting video material of a wide variety of mechanical movements, including drawers of CD players/changers, a mechanical gripper at a fancy fair, train doors and an ATM returning a card. As students try to categorize the movement videos they develop a feel for the qualities of movement. For example, CD trays do not move linearly but accelerate and decelerate, cassette compartments turn out to be damped, and CD changers have compound movements (translation + rotation).

3.2 Moodboards

Moodboards are collages of images expressing emotions through form, colour and composition. In the student project, for the most part, the moodboards were useful in exploring expressiveness of appearance in two dimensions, thus providing a stepping stone towards expressiveness of appearance in three dimensions.

3.3 Sketching

Sketching proved useful both for expressiveness of appearance and as a first start towards expressiveness of movement (here we mean sketching in two dimensions). For the telephone cradle sketches proved useful to explore the object-character of possible telephone cradles and to discover or reject certain metaphors in an early phase (for example sketches of the same object in active or a sleep-state). Some students sketched cartoon-like storyboards, exploring stances that their product would need to move between.

3.4 Pre-modelling

Pre-modelling is essentially 3D sketching. Models are made from cardboard and foam (mostly using a glue gun). Pre-models offer a way to explore both expressiveness of appearance and of movement. Some of these pre-models were expressive in a static, sculptural manner, based on the assumption that it useful to explore the relation between form and emotions before adding movements (Figure 1 and 2).

Figure 1. Exploring form and expressed emotions (1).

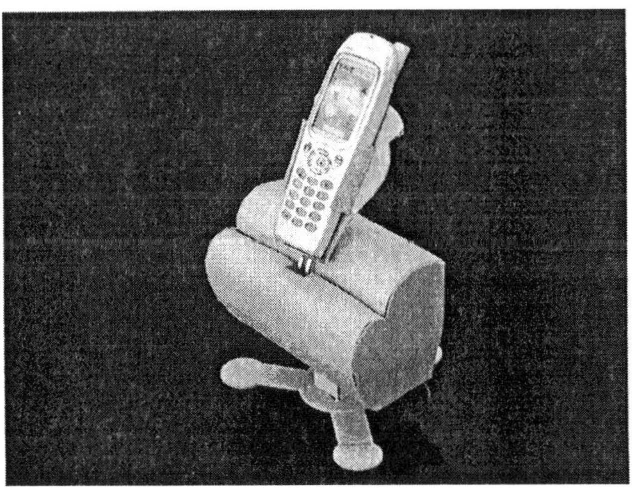

Figure 2. Exploring form and expressed emotions (2).

The cradles of Figures 1 and 2 were intended to convey urgency for negative and positive valences, respectively. Some preliminary investigations revealed that the valence seems to work as intended but the arousal did not (it was reported that the cartoon-like form was associated with "not serious" and "not business-like"). Although no full experiment was done yet, we found emotion self-reports using a paper and pencil version of the Self-Assessment Manikin (SAM; Lang 980) useful.

Figure 3. Exploring form and expressed emotions (3).

The pre-models of Figure 3 were used find out whether there was a difference in expressing anger and panic.

Other pre-models already explored qualities of movement. One student added an excentric weight to a hand-sized egg-shaped foammodel, leading to gentle, curvy and soothing movements.

3.5 Acting out (full body movements)

We asked students to act out emotions. Video-taping these miniature 'emotional plays' allowed students to analyze and discuss their movements and lent focus and concentration to the session. Acting out is familiar from Burns et al. [3]. In our case, the students acted out not so much the interaction between user, machine and environment but actually stood in for the machine and acted out how it would behave physically. The technique also bears resemblance to that used by Buur et al. [personal communication], though the emphasis is on the emotional content of movements rather than on their interaction style. Initially, we were somewhat hesitant to offer students this technique as we were concerned that this would lead students on an anthropomorphic route, whilst we were aiming for products that had expression appropriate to products.

3.6 Acting out (hand movements)

Somehow the students were aware of this too, and by their own accord they added this technique which is a modification of acting out full-body style. Instead of acting out emotions with their whole bodies, they used only their hands, again recording their movements on video tape. Excluding facial expressions already heightened the level of abstraction.

3.7 4D sketching

In the next section, we describe a technique developed by the second author which we propose to add to the design process of interactive products to make possible sketching in the temporal dimension. This technique moves beyond on-screen and wizard of oz simulations to physical prototypes with autonomous behaviour. The second author first applied the technique in 'phone robots' of Kyffin and his design and later the technique was explained to our students who used it to further explore the expressiveness of their 'walnut machines'.

4. SKETCHING IN THE TEMPORAL DIMENSION

In addition to the techniques of "acting out" at the full-body level and at the hands level, we propose to bring 3D models to life by providing them with actuators, sensors, and processing power. Such augmented cardboard and foam pre-models, could allow designers to explore phyical autonomous movements in a sketchy manner. The actuators, of course, are indispensible to create motion. The sensors are needed to add interactivity or to serve as input when mediating motion. Processing is necessary to control the sequencing and the detailed timing of the actions as well as the rules that connect actuators to sensors. We discuss the actuators first (this section is more technical than the others because the technicalities matter when one must choose among alternative actuators). Although, in principle, there are no limitations to the power and the sophistication of the actuators that could be used, we restrict the discussion here to the types we have been using in at least one of our 4D sketching activities. These are:

- solenoids
- DC motors
- stepping motors
- servo motors

Solenoids are compact actuators that convert electrical energy to linear mechanical action. They contain a coil and a steel plunger that is drawn into the coil's magnetic field when energized. They have three disadvantages: their force is not constant over the stroke; most types require return springs, and they tend to get hot when activated for more than a few minutes (we once almost ruined a polystyrene foam model of an active Pacman-like game-support character because its state after the game-session was left undefined).

DC motors are very cheap, small, efficient and silent but an unless a gearbox is fitted, they rotate too fast for most applications (e.g. 3000 rpm). The torque (rotational force) is inversely proportional to the speed of the output shaft. The gearbox makes them larger and noisier. Controlling their speed is not just a matter of controlling the DC Voltage; a pulse-width modulation is required instead. Controlling their position requires some kind of feedback (which is precisely what a servo-motor does).

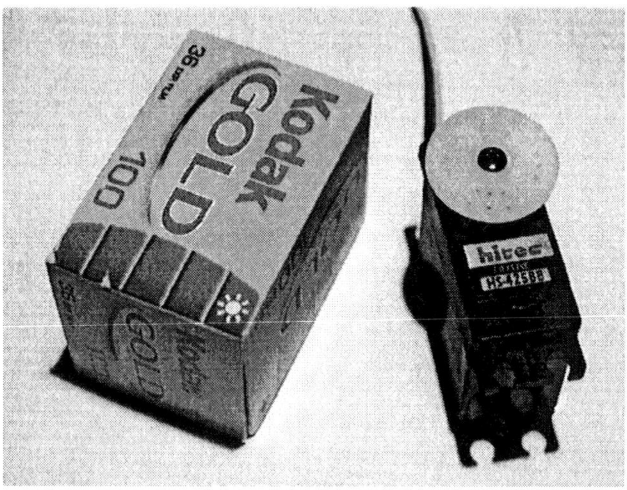

Figure 4. Servo motor

Stepping motors are best understood as electric motors without commutators. The task of the commutators is shifted to the external motor controller, typically the processor plus some extra circuitry. The motor can be held in any fixed position as well as being rotated one way or the other. When a stepping motor is instructed to go to a certain position but the control loop is overtorqued, the motor must be reinitialised and instructed again (this is different for servomotors).

Servo motors contain a DC motor and a variable resistor measuring the rotor's position. These are mechanically and electrically configured to form a feed-back system: the motor will apply more force when encountering resistance and it will try to return to the desired position after a disturbance.

A small gearbox is built-in. They are not cheap, however (the ones we used cost 40$ each). Their control commands are in the form of repeated pulses whose width indicates the desired position (from 1.25 to 1.75 millisecond). Typically they can turn 210°. This is the preferred actuator for our 4D sketching approach.

Several types of sensors have been used: passive infrared (detecting the heat of humans), active infrared (detecting a reflected signal), and capacitive touch sensors (based on QT310 QProx™ programmable capacitive IC). These sensors are smaller than the actuators and as such pose no space problems in the sketching. The capacitive touch sensors demand short wires, however; otherwise the wires would act as capacitors themselves (so for a sketch having four touch sensors, a small board with four 8-pin chips had to be housed inside the model).

Figure 5. PIC microprocessor board

Processing power is provided by a microprocessor board containing a 20Mhz PIC processor to be programmed in JAL. (Jet Another Language). Although other processors could be used as well, such as the well-know BASIC stamp, we found PIC convenient because it had been introduced already into our Industrial Design teaching program by our colleagues of the TU/e faculty of Electrical Engineering and also because it has more than 20 configurable input-output pins. The programs can be created using any PC or laptop computer as the JAL language is open source and cross platform, available on SourceForge. After program compilation and download, the PIC board works in a stand-alone fashion. Downloading is through a protocol called WISP that uses the serial port. Although, in principle, the microprocessor board could be housed inside a model, until now we found it convenient to hide it just in a drawer of the demo-table or inside an exhibition console. The processor is reliable and demands no login or boot procedures (it is "turn-key").

The JAL language has Pascal-like programming constructs. It allows direct access to the input-output pins (this is an advantage when compared to higher-level languages such as Java, where the language's security mechanisms implies that hardware is accessed through special interface packages). The data structures are primitive, however: bits and bytes only. To give an impression we show the program which makes a LED connected to pin a0 blink.

```
include f675_20
include jlib

var bit LED is pin_a0
pin_a0_direction = output

forever loop
     delay_100ms( 4 )
     LED = high
     delay_100ms( 4 )
     LED = low
end loop
```

In a typical sketching application with two servo-motors, the program contains two sub-programs which represent finite-state machines, one for each servo motor. Each finite state machine inspects the relevant sensor pins, inspects some variables, makes some decisions, and writes the desired position for its servo-motor to a special variable. There is one main control loop, which invokes both finite state machines. Translating the position value into input output behaviour with delays (in the 1.25 – 1.75 millisecond range) for the servo-motors is done in one main loop. The approach resembles the synchronous programming style often used for so-called programming logic controllers (PLCs). The advantage is that the program looks like a set of nicely structured parallel processes, but no multi-tasking operating system is needed. The turn-around time of a typical edit-compile-run cycle is one or two minutes. In our view, this is still fast enough to call the approach "sketching".

During the telephone cradle and walnut vending machine projects we saw variations of 4D sketching emerge, suited to different phases of the design process. Early on in the process, the servos are combined with quick and dirty pre-models. In this phase, the 3D configuration, the mechanisms and the form details as well as the software, which determines the behaviour, are still open for rapid modifications (Figures 6 and 7). We think of this use of 4D sketching as explorative. Towards the end of the process, the same technology is embedded in more polished and robust models. In this phase, the 3D configuration and mechanism are already 'frozen': they cannot be easily altered any longer. Of course, some of the subtle details of the behaviour such as timing can still be tweaked. We think of this use of 4D sketching as presentational.

An analogy can be drawn with 2D sketching: compared to renderings, rough sketches and presentation sketches are both forms of sketching yet each has its own place in the design process. The former to develop ones ideas on his own or perhaps in a team, the latter to present to managers or clients.

These two phases of 4D sketching are captured in Figure 8. Whilst the explorative version of 4D sketching allows the designer to carry out both a 'change 3D' and 'change behaviour' loop, with the presentational version the 'change 3D' loop is no longer an option: it is only possible to change the behaviour. Most importantly, the 4D sketching method offers a smooth transition from 'explorative' to 'presentational': the components and workflow remain the same.

Figure 6. 4D sketching explorative(1)

Figure 7. 4D sketching explorative(2)

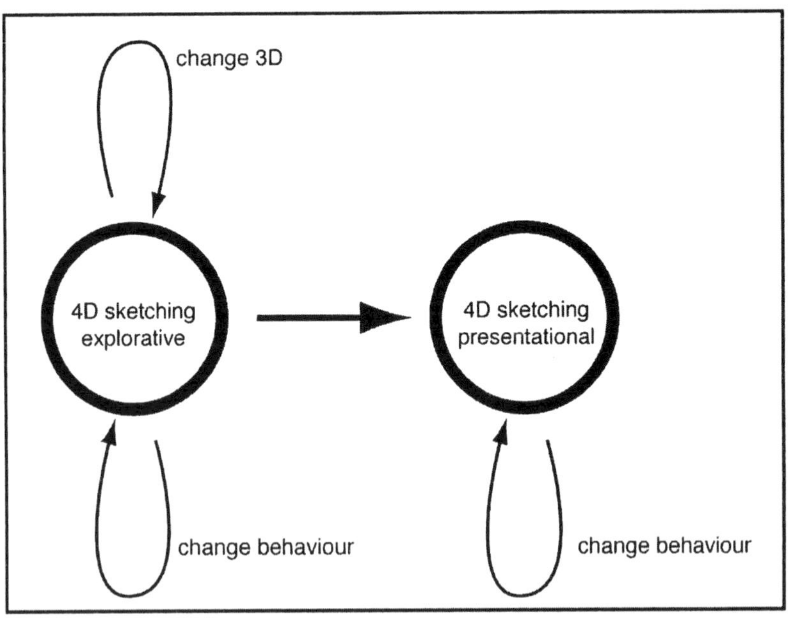

Figure 8. 4D sketching explorative(3)

5. EXAMPLES AND EXPERIENCES

5.1 Phone cradles

The 4D sketching technique has been used in the (ongoing) telephone cradle project to explore whether an active cradle can express emotions. The 4D sketch of Figure 9 is quite capable of expressing "joy" through wobbling (it is kind of top-heavy and therefore the base wobbles too).

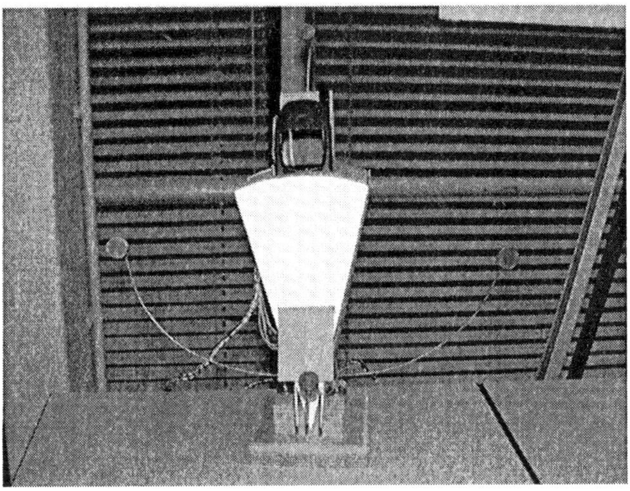

Figure 9. Active telephone cradle with wobble movement.

The models of Figures 9 and 10 have also been used to explore interaction. The model of Figure 9 has four touch-sensors. In each case the phone will move towards the user touching a sensor. Each the cradle has a lift; so when in Figure 6 the upper touch sensor is touched, the phone comes out whereas when the lower touch sensor is touched the phone moves back inside the cradle. The other model, shown in Figure 7, has just one sensor for the lift instead of two; the question we explored was whether users would recognize the "logic" of this second model , where the sensor works as a toggle switch (most test subjects had no difficulty with guessing or grasping the toggle behaviour). But it also became clear that involving test subjects demands a clear context or useful application.

Therefore in the student projects, clear choices for the context were made by us from the very beginning. The object was to perform a simple but useful function: to offer a little food. Then the students decided that this was to be a walnut.

Figure 10. Active telephone cradle with twist movement

Here we describe the design rationale and the characteristics of the five walnut machines.

Figure 11. Crack, the angry walnut machine (closed)

5.2 Crack (anger) -design: Rombout Frieling

Frieling describes anger as a contrast between inside and outside, of initially cropped up inner tension against the outside world. Anger may change one's appearance rapidly and unpredictably, through fast and powerful motions in a trend of escalation. At the climax of one's anger, there may be a kind of anger overflow.

Frieling translated this description into 'Crack', a spherical machine whose shell consists of five blades with serrated edges. Initially, Crack's shell is fully closed with only the blood red, serrated spaces between its blades indicating its unrest. When approached, Crack lashes out at its user through increasingly large movements of one randomly chosen blade at a time. Finally, Crack totally cracks open into an aggressive angular, blood-red shape and ejects the walnut vertically.

Figure 12. Crack, the angry walnut machine (open).

5.3 Dolly (love) - design: Jan Hoefnagels

In showing love or affection, one can do anything from comforting, via playing with, to teasing the loved one. Hoefnagels translated love into a machine called Dolly which has a wooden, smoothly polished, organic body. Dolly exhibits smooth and slow movements, intended to resemble a comforting caress. The movements curve and turn and topple past the user, giving an impression of teasing, of playing with or nuzzling up to him or her. Finally, eight bead-terminated pins slowly expand from Dolly's main body, offering the user the walnut.

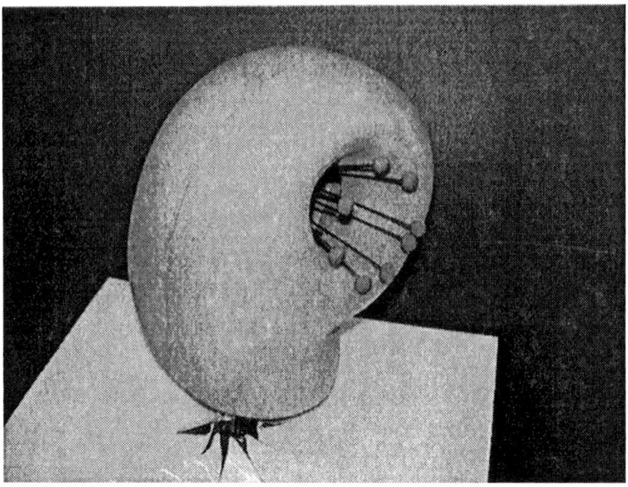

Figure 13. Dolly, the walnut machine in-love (nut still inside)

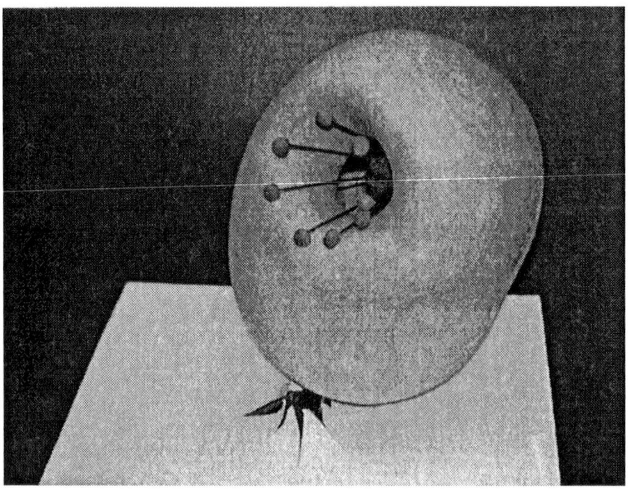

Figure 14. Dolly, the walnut machine in love (nut coming out)

5.4 More Walnut machines

Emotion: *Enthusiasm*. Ralph Zoontjes' model is used to explore expressing the emotion enthusiasm. His description of this emotion contains several stages. First, there is a suspense stage in which one expects an event that would be very advantageous. Second, one becomes more active and interested, concentrating on whether the event is going to happen or not.

Then, as it is very likely for the event to happen, the excitement builds up, one start moving more, thus letting out some of the energy that has gathered inside. Finally, the real enthusiasm bursts out. From various observations (people watching football matches, happy dogs), Zoontjes concluded that there are many movement with high frequency and low amplitude to begin with but that enthusiasm often ends with very intense, large movements. Zoontjes implemented these ideas into Phunx, a bright red and green coloured machine of organic form. Initially, Phunx slowly 'wags' its red top. When approached, it starts wagging faster. Finally, it lifts its white 'walnut scoop' in three quick, increasingly large movements, leading to the walnut being launched in the user's direction.

Emotion: Panic. Dirk Volman noted the following dynamic characteristics of panic. Firstly, one's movements spring from a single point. Secondly, the movements lack focus, because one wants to do things all at once, worsening one's panic only further. And thirdly, a movement creates a counter-movement because one wants to do everything at once and because one doubts the initial move was effective. And each counter movement creates a counter movement which creates a counter movement etc. etc. Finally, one startles again as one realizes one's problems are solved after all. Volman translated these characteristics into a semi-spherical machine called Caos. Caos violently moves random sections of its shell outwards and retracts them again, to finally extend all of them simultaneously. The nut rolls out in a rather uncontrolled manner and, after a short break, Caos retracts all its moving parts and comes to rest.

Emotion: Shyness. Melika El-Massoudi noted that shy people first turn away from the requesting user, often shielding their face with one hand, blushing and hesitating between doing nothing and fulfilling the request. Her machine consists of two transparent semi-spherical plastic shells in whose center the walnut lies. As the user approaches the machine, the shells turn their back on the user, showing a different colour. They then move between shielding and exposing the walnut to, finally, fully expose it to the user.

5.5 General observations

On the basis of the given examples and experiences, a number of general observations can be made. First, it is absolutely worthwhile to create models which exist in the physical dimensions of the real world, not just 3D simulated in Virtual Reality. The point is that the interaction takes place by the user's real body in a multi-modal way. The sensors are not abstract buttons: the user really feels the touch button and the user notes that the IR sensors, depending on their type, do or do not have a directional sensitivity. Moreover, the sensors may move themselves relative to the phone or the nut.

Secondly, it became clear that form, colour, motion and even sound are hard to separate. In Crack and Caos these were designed to re-enforce each other (the snapping sound of solenoids fits well with panic); the surprise-effect this had on test subjects was enormous.

Thirdly, we found that a well-defined *context* was indispensable when entering a rational dialogue with a user or a test-subject. On the other hand, even without proper context, observers were intrigued by any interaction behaviour or autonomous behaviour of the models, except for simple and predictable behaviour.

Finally, we found that there are basic laws of object motion that emerge during 4D sketching but that are easily overlooked when working in 2 or 3D. In particular, any finite translation or any rotation of less than 360° has sooner or later to be followed by a movement in the opposite direction. For example, the blades of Crack move outward very fast, as a part of its expression of anger; but when folding-in gently, they weaken the anger. But closing fast again makes the expansion and the contraction symmetric, which is also not the designer's intention (for anger one needs the aggressive expansion). This type of problem continues to exist when there are several parallel, independent finite translations or limited rotations (like the four blades). But it is possible to find ways out of this dilemma when two movements are coupled (like the rotation and translation of Dolly).

6. RELATED WORK

In recent years there has been a growing interest in toolkits which facilitate the building of physical, mechatronic prototypes, such as Phidgets (Greenberg & Boyle, 2002), iStuff (Ballagas et al., 2003) and the Calder Toolkit (Lee et al., 2004). What these toolkits have in common is that they intend to make the creation of electronic prototypes possible for non-experts, i.e. those not skilled in electronic engineering or information science. Frequently, an analogy is drawn with graphical user interfaces in which GUI toolkits with drag and drop widgets that come with their own behaviour have made possible the development of GUI applications by non-programmers. All of these toolkits make use of physical input and output building blocks that can be linked to an application running on a PC. The main idea behind this approach is building on existing knowledge: by using a PC the user can program behaviour in a graphical environment using high level languages such as Macromedia Director's Lingo, Visual Basic or Java that many in the interaction design community are already familiar with.

In contrast with these toolkits, our 4D sketching method does not rely on such tight integration between electronics hardware and the PC. The hardware components are not represented in anyway in the graphical environment and do not require drivers. Instead, the PC is merely used to type in JAL code and download it to PIC. We deliberately do not use a 'PIC simulator' on the PC: all exploration is done through rapid iteration cycles, comparing the behaviour in the physical world with the designer's expectations. Actually, the exploration process starts at the beginning of the project when a working I/O program together with a sensor and servo is delivered to the students. From that point on, the process is a matter of modification, re-use and thus exploration. The combination of the JAL language and the PIC microcontroller offers direct control over input and output ports in one or two lines of code. The JAL language is easy to step into and has a gradual learning curve, just like Basic. Typically, JAL leads to small, single file programs that are easy to read and encourage re-use of software.

Compared to other microcontrollers with easy to learn languages such as the Parallax BasicStamp and Javelin, the PIC microncontroller is very cheap. The high spec 16f877 that we use here costs around 10 USD. A comparable BasicStamp or Javelin costs about five to nine times as much. The low cost of the PIC not only makes it less costly to create multiple prototypes, it also means that we can experiment more freely: a couple of burned microcontrollers won't break the bank.

Finally, Microchip, the makers of the PIC microcontroller, offer a path for further miniaturization. Whilst our programming board makes use of relatively large 40 pin DIL packages, the JAL code will run 'as is' using the same input/output pins on a miniature surface mount version of the same controller.

7. EVALUATION AND OUTLOOK

We described how we and our students made the leap towards sketching in the temporal dimension. As we are based at a technical university, we are interested in letting students gain proficiency in integrating technology in a practical, hands-on manner, thus creating a new designer 'breed' that can move beyond static models and on-screen simulations towards working, mechatronic prototypes.

It was possible to “package” the 4D sketching approach by instructing the students on the acting-out techniques (Sect. 3.4) and by providing them with the materials shown in Figures 4 and 5 as a part of a two-our demo/instruction session (the default was two servos, a PIC and a sample program; other actuators and sensors were dependent on each student's own initiative). Once the technicalities of 4D sketching are mastered, creating a new model from scratch takes less than four hours. Modifying an existing model for exploration purposes is a matter of minutes.

Let us return to the question from the introduction: How can the form, empowered with behaviour, communicate human values, culture values, societal values, tribal, community, brand values? It has become clear that objects can express at least basic emotions through movement. The 4D sketching technique is useful as a tool to explore basic emotional expressions and at least promising to explore the next levels of complexity, mediating values.

ACKNOWLEDGMENTS

We gratefully acknowledge the enthusiastic contributions of our students Rombout Frieling, Jan Hoefnagels, Raplph Zoontjes, Dirk Volman, and Melika El-Massoudi, the technical support of Ton van der Graft and the help and cooperation offered by Kees Overbeeke and Joep Frens.

REFERENCES

Aldersey-Williams, H., Wild, L. Boles, D., McCoy, K., McCoy, M., Slade, R., & Diffrient, N. (1990). The New Cranbrook Design Discourse. New York: Rizzoli International Publications.

Avrahami, D., & Hudson, S.E. (2002) Forming interactivity: a tool for rapid prototyping of physical interactive products. Proceedings of DIS2002, London, pp. 141-146Ballagas, R., Ringel, M., Stone, M., & Borchers, J. (2003). iStuff: a Physical User Interface Toolkit for Ubiquitous Computing Environments. Proceedings of CHI2003, pp. 537-544.

Burns, C; Dishman, E; Verplank, W. and Lassiter, B, (1994) Actors,Hairdos &Videotape – Informance Design. In: CHI '94 Conference Companion, ACM, Boston.

Greenberg, S., & Boyle, M. (2002). Customizable physical interfaces for interacting with conventional applications. Video Proceedings of the ACM UIST 2002 15th Annual ACM Symposium on User Interface Software and Technology. ACM Press.

Hummels, C.C.M., Djajadiningrat, J.P., & Overbeeke, C.J. (2001). Knowing, doing and feeling: communicating with your digital products. Interdisziplinäres Kolleg Kognitions- und Neurowissenschaften, Günne am Möhnesee, March 2-9 2001, 289-308.

JAL open source project: https://sourceforge.net/projects/jal/

Lee, J.C., Avrahami, D., Hudson, S.E., Forlizzi, J., Dietz, P.H., & Leigh, D. (2004). To appear in the proceedings of DIS2004.

Krippendorff, K., & Butter, R. (1984). Product semantics: Exploring the symbolic qualities of form. Innovation. The Journal of the Industrial Designers Society of America, pp.4-9.

RE-CONCEPTUALISING 'FUN'

THROUGH VIEWER'S EXPERIENCES TO BUILD NEW HOME SYSTEM INTERFACES

Greg Hale
CUHTec, University of York, U.K.

Abstract: This paper reports work from the first stage of a project investigating movie based fun, focused on content and delivery by broadband into the home and to mobile telephones. After considering 'fun', the changing technical and social context of digital technology and digital experiences is examined, to understand how people's enjoyable experiences are changing. Then a detailed scenario of what is here called the experience rich home interface is presented, before highlighting some of the design issues. The paper concludes by identifying the next phase of the research, an engagement with people's experience of movie content.

Key words: Home systems, fun, interfaces.

1. INTRODUCTION

This paper is focused on people's experiences of digital content and delivery, including people's movie experiences (as opposed to the movie design languages of cinematography and editing) as a source of challenge, critique, and design ideas for new interfaces. The paper forms part of the work for the *'Qualitative and quantitative studies of fun with movie based entertainment via mobile telephones and websites'* project. After outlining a high level conceptualization of fun, the paper considers the changing nature of digital experiences and movie experiences. The implications for new interfaces are then illustrated in a scenario, drawing on themes from the preceding section to describe a new type of experience rich home system and its interface. Following this, some design issues of relevance to this experience rich interface are highlighted. The paper concludes by identifying the next phase of the research, an engagement with people's experience of movie content.

2. FUN

The study of 'fun' is relatively new, though there have been substantial research contributions in recent years (e.g. Blythe *et al.* 2003; Green and Jordan 2002). The growing research interest in fun - particularly as it pertains to interactive devices - is unsurprising given the social significance of interactive fun, the cost of creating interactive products which supply fun and the interests of media producers in improving and extending the fun provided in their interactive products.

Research into 'fun' is complicated by the fact that many terms are used almost interchangeably for fun, such as enjoyment or pleasure (Blythe and Hassenzahl 2003). The origin of the word is uncertain, with the usage changing radically over time (OED Online, Webster, WordNet). The word may be of Celtic origin, from 'fonn', meaning pleasure, but the earliest recorded usage (1685) was to cheat or fool someone. The change to a modern usage came in the eighteenth century and may be linked to the industrial revolution, with its rationalization of work through mechanization. This could have resulted in 'fun', with it's emphasis on frivolity and the trivial, becoming seen as the opposite of 'work' (Blythe and Hassenzahl 2003). In modern use 'fun' refers to enjoyment and amusement, usually in a recreational setting: a positive internal psychological state and an external facilitating context.

Much of the research on fun has focused on interactive systems such as computer games (e.g. Carini 2003; Federoff 2002; Kim *et al.* 1999), a medium known for providing fun intense experiences. Films also provide intense experiences, yet research linking the movies to human-computer interaction has tended to focus on cinematographic and editing conventions as a means of improving existing interfaces (e.g. McKendree and Mateer, 1991; May and Barnard 1995).

Fun therefore has as a minimum the following global characteristics:

- Recreational setting
- A positive internal and interpreted bodily/psychological state (for example, enjoyment)
- External facilitating context

In the section that follows, changes in the external facilitating context and recreational setting are focused on. These changes are likely to lead to changes in the interpreted internal state. How this happens, in relation to the movies, is part of the ongoing programme of research to be undertaken in the next phase of the research. Also part of the ongoing research is an exploration, in the context of movies, of more local (i.e. lower level) elements of fun, which might lead to designers designing *for* experiences (McCarthy and Wright 2004).

3. CHANGING DIGITAL EXPERIENCES AND CONTEXTS

As more people in the industrialized nations become part of the digital networked economy (BT 2004) people's experiences of technology – how they use and interact both with media content and the delivery mechanisms that bring this media content into their lives - are being altered by pervasive changes in the digital context. In the future, it will not be technology *per se* but digital experiences which are available anytime, anywhere. These experiences will interpenetrate both physical reality and peoples 'felt' personal lives.

At the outset, it is important to say that the strong claim that experiences can be totally designed is not being made in this paper- people vary dramatically in their response to identical stimuli and they are active in construing meaning. But experiences can be 'designed for' (e.g. Fiore 2004). The position taken in this paper is that there are both wide individual variations and also central commonalities in people's construal of experiences.

In this section I explore how people's experiences of digital technology are changing, seeking to identify changes taking place now which may in the future shape what I call the hyper-digital age where digital experiences are constantly available and personally meaningful and influential in people's lives. The changes include increased exposure to interactive media, the growth of mobility, the digitization of the movie industry, the shift in locale for digital experiences, the development of consumers as controllers of content and the growing research interest in experiences.

The argument of this paper is that these changes in the external digital context are leading to changes in the internal experiences that people have- and that both these internal and external changes provide a new context for new types of interface, in particular what I call experience-rich interfaces.

3.1 First trend: increased use of interactive entertainment, the Internet and personal computers

The first trend in the changing landscape of digital experiences is the increased use of interactive entertainment, the internet and personal computers. Ofcom report that 64% of adults in the U.K. have a PC at home, with 53% connected to the internet, and 25% connecting via broadband, up from 16% over the past year (Ofcom 2004d). Uptake and usage has also risen: for the period 1999 to 2003 time online has increased from two hours via dial-up in 1999 to 16 hours a week in broadband households (Ofcom 2004a). DVD player ownership in Europe has almost doubled from 30 million in 2002 to 56 million in 2003 and is forecast to reach 67% of European homes by 2007 (Dawtrey 2004). Ownership of digital televisions currently stands at around 55% of all UK households (Ofcom 2004a) with market-led adoption of digital TV predicted to reach 78% by the cut-off date set by the government for switching off analogue television (Ofcom 2004b) – there will doubtless be a decisive shift to digital as analogue options run out in 2010.

There is also growth in people's experiences of interactive entertainment, directly affected by the growth in popularity (and notoriety due to sexual and violent content, in some cases) of computer and video games. In 2003 the global leisure software market was valued at $18.2 billion, an increase of 10.8% from 2002 (Elspa 2004). In the UK, the market in 2003 was estimated as being worth £1,152 million, an increase of almost seven percent from 2002, with the UK's leisure software market having grown by over 100% from 1997 to 2003 (*ibid.*). This compares with cinema box office growth for the same period at 30% and video/DVD rental for the same period at 14% (*ibid.*), though these figures may be understated due to people illegally downloading pirated film content (see Cole 2004).

Not all digital experiences are showing such clear trends – the prognosis for broadband delivery of television content is unclear. Current indications from other countries such as early adopter South Korea are that broadband television is used largely for specialist programs (Ward 2004). In the UK there are currently over 10 million homes with some form of on-demand television programming. This is predicted to rise to 15 million by 2008 (BBC 2004), with movies being mooted as a central driver of this trend (*ibid.*). Though the number of home with Internet Protocol Television (IPTV) is predicted to rise to 4.5 million in Europe by 2008, currently such services are struggling financially (*ibid.*). It is too early to say for certain whether such media will become pervasive (Anderson 2004) though these technologies will certainly carve out their own specialized areas of use – and there are significant business entities seeking to make such technologies pervasive (Gates 2004). It may also be noteworthy that time spent viewing television has increased only marginally, from 25.6 hours a week in 1999 to 26.1 hours in 2003 (Ofcom 2004c), which suggests that leisure time is being shifted from television to other activities.

This first trend means that users of digital media are gaining increased familiarity with a constantly expanding multiplicity of interface and device control structures, interactive links to advertising content and a multiplicity of intense digital media experiences.

3.2 Second trend: the growth of mobile telephony

The second trend in the changing landscape of digital experiences is the growth of mobile telephony, one of the most influential personal technologies in the UK. Currently 86% of households in the U.K. own at least one mobile telephone (Ofcom 2004c). As more applications are bundled in and cost structures become more acceptable, so third generation (3G) mobiles and services will become increasingly important in many peoples lives, as is the case with earlier model mobile telephones (e.g. Peters 2002). With 3G mobiles slowly appearing (Durman 2004) people will eventually be presented with an ever wider smorgasbord of digital experiences, facilitated by two 3G technical developments. Firstly, 3G provides more bandwidth (always-on broadband connection at speeds of 144Kb/s to 2Mb/s). Secondly, display screens are constantly improving, with high-quality colour displays now standard. These technical advances facilitate the enjoyment of services such as fax, email, web access, videoconferencing and movie-related content such as trailers or tie-in competitions, music videos with high quality audio and multi-player mobile gaming – Bill Gate's view of "digital entertainment, everywhere" (Gates 2004). Uptake of 3G currently is slow, with services still being rolled out in time for Christmas 2004. Billing structures are also not yet finalised, which will hinder uptake because consumers worry about the cost of their mobile calls (The Work Foundation 2003). In addition, technical issues remain - 3G networks are sparse and growth is likely to be slow, with London and the South of England favoured (Durman 2004).

Currently uptake of 3G is projected to be not more than 20% by 2007, but the development of high value innovative services and applications could exponentially boost these figures. With time on mobile phone calls up from 10 minutes in 1999 to 27 minutes a week in 2003 (Ofcom 2004c) mobile telephony is potentially one of the most important delivery technologies for new digital experiences. In the long term mobile telephony will develop greater capacity to bring high-quality audio-visual content to users.

This second trend means that people will increasingly become accustomed to and want rich sensually and emotionally satisfying digital experiences, anytime, anywhere.

3.3 Third trend: the movie industry goes digital

The third trend in the changing landscape of digital experiences is the shift of the movie industry from analogue (celluloid) to digital. This is coming about as movies are entering their own digital age as digital goes beyond digital manipulation for special effects to encompass production and distribution.

Digital cinema (d-cinema or e-cinema) is set to revolutionise distribution and exhibition by removing the cost of celluloid prints from the distribution equation. In the UK, d-cinema is making progress, with the Film Council seeking to install 250 digital screens in the UK, albeit in 'art house' cinemas (UK Film Council 2004). Other countries have similar plans, though there are still digital standards and cost issues to be resolved (Shackleton 2004, Lall 2004, Paquet 2004). Distribution and exhibition is also being affected by the internet, which is slowly becoming a significant medium for the distribution of movies generally, with established firms working to create viable delivery models (Olsen 2004).

The development of d-cinema is being significantly helped by the development of high-definition video cameras capable of producing images of sufficient definition for cinematic exhibition. High definition video cameras were used most famously by George Lucas in *The Phantom Menace* (used because of the effects-heavy nature of the film). This gave a significant boost to the visibility, credibility and perceived viability of the use of high-definition video for movies intended for theatrical distribution. The growth of high-definition video movie production is also being promoted by the insistence of some funders such as the UK's Film Council that entry level filmmakers use it for low-budget productions.

3.4 Fourth trend: digital experiences are changing locale

The drive of media companies to market their products more effectively and to generate multiple streams of revenue is having a profound effect on the locale of digital experiences. Increasingly integrated media experiences across multiple channels are offered to consumers, via newspapers, websites, mobile content, TV/radio etc., even interacting with the physical world of consumers 'real' lives. So, for instance, 'Uncle Roy All Around You', is an internet based game where you hunt for 'Uncle Roy' in the real world (http://www.uncleroyallaroundyou.co.uk/). More widely known is the Blair

Witch project, where part of the initial public interest was the belief that the website was giving 'real-life' updates on real-life events. This emphasis on media products and experiences spilling over into real life is likely to be heightened by the development of location aware mobile telephones which can be used as the basis for new multi-player games such as electronic shoot-em-ups played out in real life (Norris 2003). And availability of digital entertainment is being widened by the current Microsoft 'Digital Entertainment Anywhere' initiative (Gates 2004).

The audio-visual experience of cinemas is also shifting. Increasingly, leaving aside the social aspects of cinema, the audio-visual aspects of the cinema experience are available in the home as 'home cinema' technologies provide higher quality audio and picture than traditional television sets.

This fourth trend means that providers will find it easier to integrate movie content with other content, as Vodafone are currently doing (Durman 2004), to provide sensually and emotionally rich digital experiences based on characters and narratives.

3.5 Fifth trend: the consumer as controller of content

The fifth trend is the consumer as controller of content, which relates to the democratization of media production through experiences of media control in the DVD experience and the availability of low-cost computer assisted media production. One output of this trend is the re-use and personalization of existing media content (Bolter and Grusin 1999), shown most clearly in the production of short movies by fans, involving pastiche, spoof and new contexts for familiar media offerings (e.g. *Atomfilms.shockwave.com* and *ifilms.com* for example; see Jenkins (undated) for an in-depth consideration of these phenomena). Ultimately, personalizing content after delivery may lead to personalizing media content during delivery. Currently there is some movement towards this model, with the European Union's 'New Media for a New Millennium' initiative aiming to develop the tools for personalized interactive stories (including movies), adapted 'on the fly' for viewers (New Media for a New Millenium 2004).

It remains to be seen whether this New Media initiative adequately recognizes that certain media experiences such as watching new releases of movies are strongly socialized - a significant aspect of the pleasure is the ability to discuss the same non-personalised movie with friends. It is also possible of course that new forms of personalized movie experience could lead to new topics of conversation and new and attractive pleasures. So, for instance, it may be possible that in twenty years time one form of movie experience will involve groups of people watching a downloaded movie at the same time, with sound feedback from all audience members (or designated 'buddies'), to facilitate the social experience during the movie (i.e. shared laughter) together with video-conferencing and interactive games after the movie to extend the experience.

This trend means that viewers can and increasingly will actively reconstitute media content according to their preferences and desires, whether just for fun or as an expression of personal creativity or for other reasons. Whatever the pleasures of reshaping media into new forms for the average viewer, most of the amateur efforts available from internet sites are severely lacking in narrative drive and understanding of the fundamental techniques of movie making.

3.6 Sixth trend: growing research interest in 'felt life' approaches to experience

One recent approach to digital experience that roots itself decisively in people's felt life has been put forward by McCarthy and Wright (2004). They seek to go beyond accounts of people's use of technology as described by usability and it's dimensions and also beyond accounts of users as cogs in virtual machines, as social actors and as consumers. Instead, McCarthy and Wright focus on interactive technology in the context of people's felt life - an account rooted in emotional, sensual and aesthetic aspects of interaction. They identify four 'threads' of experience – the sensual, the emotional, the compositional and the spatio-temporal. Here the method is not definitional but qualitative, investigative and illustrative, seeking to capture people's felt-life accounts and their responses to the technologies they use.

This sixth trend means that theoretically grounded approaches to the study of experiences are being developed which will facilitate the investigation of people's content experiences in and across the range of new configurations of digital media that are currently being developed.

These trends represent shifts in the global elements of fun- the changes in the recreational setting and the external facilitating context. Now users are getting used to a greater number and variety of fun experiences (Trend One: *Increased use of interactive entertainment, the internet and personal computers)* with new forms of entertainment available anytime anywhere (Trend Two: *the growth of mobile telephony;* Trend Four: *Digital experiences are changing locale*). There are also shifts in the locus of control in the production of fun with the digitization of content, particularly in relation to the movies (Trend Three: *the movie industry goes digital;* Trend Five: *the consumer as controller of content*). Finally, new research (Trend Six) into experiences opens up the possibilities of 'filling in' some of the lower-level elements of fun (without necessarily suggesting a one-to-one correspondence with visual, sonic or story elements and a fun in all viewers at all times).

So what might a new home systems interface based on fun experiences look like?

4. THE 'EXPERIENCE INTENSE' HOME SYSTEM INTERFACE- EXPLANATION AND DISCUSSION

This section of the paper seeks to show the relevance of the trends discussed in the previous section for an alternative conception of the interface - the experience rich home interface. An extended scenario is presented, to bring this concept to life. Current technology could be used to build this interface.

It should be pointed out that this conception of an interface is not reducible to a souped-up desktop theme with bells and whistles, a 3D interface or an intelligent interface (e.g. Bowen *et al.* 2002) as these are normally conceived. Equally, this interface is not just another computer game: there is a difference in the immersive experience of computer games and movies. Computer games are intensely engaging and immersive, but the immersion can be broken when errors are made and self-performance monitoring takes place. Movies, by contrast, are totally immersive as long as there are no environmental distractions.

> It was nearly six-thirty at night when Alicia Smith turned her car at last into the drive of her house. It had been a hard day's work at the canning factory where she is Manager for Creative Projects in Midtown, USA. At thirty-two she feels she has achieved all her dreams - Manager for Creative Projects, great husband, great daughter and the latest in responsive homes. Tiredly, she pushed her long red hair back from her face, longing to be out of her Armani business suit and into jeans and a t-shirt. John, her husband, won't be home until tomorrow. Alicia is thinking of watching a movie that both she and John love.
>
> The Advanced Responsive Home Unit, Design Two (ARHU-D2), the 'brains' of the responsive home, has noted that Alicia is nearing the house. In her living room (Alicia's choice of venue, only messages between her and John go through to the bedroom) the flat wall screen switches on, showing her personal messages. The ARHU has over a period of time logged and analyzed both Alicia and John's messages viewing and other life patterns, creating a detailed mapping of

preferences related to time of day and other critical context issues. In this instance, the ARHU knows that John is away. The unit also knows that Alicia usually checks her messages upon arrival at home.

As Alicia enters the house, she is greeted by the ARHU unit - Alicia has been using the 'Queen of the Ziroconian Islands' world for the past two years as her immersive home system interface (IHSI, pronounced *i-see*). The ARHU knows that Alicia will be feeling a little fed-up without John and has arranged that the Grand Vizier (an animated Laurence Fishburne-*soooo* soothing!) and the Court Jester will deliver the messages to Alicia in her throne room. The unit knows that tonight Alicia will prefer these two rather than the grouchy and feisty Lady of the Royal Chamberpots who Alicia normally prefers to deliver her messages.

"Would you like to see your messages, your Majesty?" asked the Vizier respectfully, as behind him the court Jester piled up furniture, clearly intent on performing another funny but dangerous acrobatic feat. Despite herself, Alicia couldn't help but smile as the Court Jester gave her a little wink then started climbing precariously up the wobbling tower of furniture.

"Yes, thank you, Grand Vizier", Alicia replied. Just then the little lantern by the side of the throne lit up, with the words "Titanic II: Special Offer on Edible Memorabilia" but Alicia ignored it.

The first message was from Rachel, John and Alicia's daughter. As the Vizier opened the light green scroll (Rachel's chosen colour for the outside skin of her messages) which was sealed with Rachel's own large green wax seal, he announced, "A message from Princess Rachel". At the same moment the Court Jester got to the top of the pile of furniture, spread his arms wide as if at the prow of a great ocean-going liner and shouted, "Look at me, your Majesty, I am king of the world!". Then the pile slowly tilted over and fell. The Court Jester, limbs flying in all directions, did an involuntary somersault on his way down, ending up rather shakily on his knees, then pretended this was his intent all along, "Ta da!" he cried, "The Great Alfonso does the furniture leap of death for Queen Alicia!" Alicia laughed, her feelings of loneliness lifting, remembering other times the Jester's antics had amused her. The Jester looked delighted at her laughter.

The message from Rachel was just to say that she was hoping to come over for a visit next weekend. Would her mother and father be in?

Alicia sent her own message to confirm. Then, with a single verbal command "Appointments", called up the appointments diary. In her chosen world this appeared in the form of a large leather bound book. Using the virtual keyboard Alicia logged Rachel's visit. The message from Rachel was the only message known by the ARHU to be from either family or friends.

"The envoys from Daxos are here", said the Vizier. "Would you like to see them today?"

"No", replied Alicia. The ARHU sent a message to the other players in her multi-player game to tell them she was unavailable, warning one of the players to reschedule his offer of a horse drawn carriage ride for Alicia to the Canning Factory (a real life offer that could have profound diplomatic implications for Queen Alicia's government of the Zirconian Islands, if not dealt with appropriately).

The ARHU started to warm up the bathroom, surmising that Alicia would leave the rest of her messages till later, since none of the remaining messages were from known members of the family or friends. The unit had selected the movie that John and Alicia had watched together the most often: 'You've got mail', together with three other emotionally and thematically similar films that the unit had no record of Alicia seeing before, along with trailers to help her choose which one she wanted to watch. The unit also selected six tracks of music, of the sort that Alicia usually preferred to listen to when she was alone and getting herself ready for the main event of the evening, the movie.

As Alicia showered, she smiled again to herself at the antics of the Court Jester. Maybe she would let him bring the messages next week. It would make him so happy.

When the ARHU-D2 unit had first been hooked up, she would have laughed at herself for characterizing the computer program of the Court Jester as 'he' and engaging with it as a real person. That thought would never again enter her mind.

The conventional conception of the interface has as its integrating theme the fact that interactive behaviors alter system states. Movie experiences are, according to this conception of the interface, behaviourally passive, since system states (the movie) cannot be altered by anything that viewers would normally do. Movie experiences may therefore be seen as being of low relevance to the task of designing interactive interfaces, except perhaps for surface visual design features, using cinematographic and editing conventions to improve or better understand existing interfaces (e.g. McKendree and Mateer 1991; May and Barnard 1995).

The explicit focus on fun in this interface offers an alternative to the pragmatic functional approach to interface design which has its roots in the productivity software of the work place. With home interfaces, where time and task issues take on different meanings, the way is open to deepen the use of non-functional and expressive interactions with the system interface, as will now be discussed.

The ARHU system communicates via an immersive and entertainment based interface which draws on movie experiences, with characters which truly engage (unlike many computer game characters). The ARHU uses meaningful content as the organising structure. The system is mood responsive and positive mood reinforcing. It is media referential, directly incorporating movie and other content, as well as allowing selected advertising. It also integrates narrative and character driven multi-player gaming and the intersection and interaction of these games with real life.

Simple speech recognition is used for responsive characters – a pseudo-conversational style that works on the basis of reasonable inferences about user's behaviour and simple yes-no questions as prompts for user input, to allow a movie like experience with lower required occurances of overt interaction such as using a keyboard. What is being described is not *Star Trek* but rather a system that could theoretically be built today using technology currently available. A mental model of the interface still has to be internalized by the user - if Alicia tried to initiate a real conversation then the illusion and emotional satisfaction for her would immediately be broken, since the system would be unable to respond. In other words, for all its naturalistic appearance, the interactive context of the interface is highly structured. A consequence of this is the need for subtle prompts to the user within the story world when user input cannot be recognized, to prompt users to return to an appropriate interactional style without breaking the immersive experience of the interface. The interaction is therefore structured by the narrative content and common patterns of yes/no answers to questions, within the expressive 'world'.

The system analyses the owner's life patterns. This enables it to make inferences about the owner's lifestyles and current mood, allowing optimal system and content configuration, within system limitations. The system gathers location data using sensors embedded in Alicia and John's wristwatches.

Usability has been defined as effectiveness, efficiency and satisfaction to achieve specified goals (ISO 9241-11, ISO 1998). Clearly the experience rich interface stresses the 'satisfaction' element.

The question may be asked, do we want to be looked after in such a way by such a system with such an interface? Some people will express a preference for functional interfaces and non-computer controlled mood management. And there is of course nothing to stop people choosing the latest Linux or Windows software as their immersive world. Just as there is nothing to stop the ARHU system from presenting Windows or 'Queen of the Zirconian Islands', according to its judgment of the owners mood and needs.

The experience rich interface represents the substantial deepening and extending in people's lives of the recreational setting and external facilitating context of fun. The experience rich interface links to the changes in the digital context as follows.

The particular integration of the system with the user's life and lifestyle means a much greater (though cognitively less visible) integration of the interface with the user and their life (Trend One). The ARHU demonstrates a constrained case of the shift to consumers as controllers of content, with the personalization that is available to the user (Trend Five). The shift in locale is also illustrated, both because the system can be distributed for users to interact with mobile telephones and other devices away from the home (Trends Two and Four, not illustrated in this scenario) and because of the incorporation of digital movie content into the interface (Trend Three).

5. CONCLUSION AND FURTHER RESEARCH

This paper has explored the changes in the external digital context and suggested that these changes are leading in turn to changes in the fun experiences that people have. It has been suggested that these changes provide a new context for a new type of interface, called here the experience-rich interface.

The next phase of the research will examine the psychology of movie experiences more closely, primarily focused on people's content experiences within a developing framework of 'movie experience intensifiers', to understand such experiences and then deploy this understanding for improved fun entertainment content and delivery via websites and mobile phones.

ACKNOWLEDGMENTS

The project is a PhD studentship funded by Microsoft Research (Cambridge, U.K.). The project is based at CUHTec, the Centre for Usable Home Technology at York University. The main supervisor is Andrew Monk. My thanks to Andrew Monk, Gordon Baxter, Siân Lindley and Peter Wright and Jo Whirton for reading this paper through.

REFERENCES

Anderson, H. (2004). Video on Demand on the Horizon. NetworkWorldFusion, (nwfusion.com), 10 October.

BBC (2004). UK Viewers Taking Control of Television. BBC News UK Edition (news.bbc.co.uk). 13 October.

Blythe, M. and Hassenzahl, M. (2003) The semantics of fun: Differentiating enjoyable experiences. In Funology (2003), pp. 91-100.

Blythe, M., Overbeeke, K., Monk, A.F., and Wright, P.C. (2003). Funology: From Usability to Enjoyment. Kluwer.

Bolter, J.D. and Grusin, R. (1999). Remediation: Understanding New Media. The MIT Press.

Bowden, R. and Kaewtrakulpong, P. and Lewin, M. (2002). Jeremiah: The Face of Computer Vision. 2nd International Symposium on Smart Graphics, ACM International Conference Proceedings Series, pp.124-128.

BT (2004) Digital Networked Economy. British Telecommunications.

Carini, C. (2003). . Constructing a Player-Centred Definition of Fun for Video Games Design. Unpublished Master's paper. Middlesex University.

Cole, G. (2004). Will Broadband Kill the Video Store? Financial Times Creative Business. 6 July, p.14.

Culpan, T. (2004). Taiwan. Will broadband give us television on tap – or is it just a fat pipe dream. Financial Times Creative Business. 30 September 2004. p.8.

Dawtrey, A. (2004). DVD Usage Soars in Europe. Variety.com. 8 March.

Dickie, M, A. (2004). China. Will broadband give us television on tap – or is it just a fat pipe dream. Financial Times Creative Business. 30 September 2004. p.9.

Durman, P. (2004). Vodafone's £20bn bet on 3G. Sunday Times Business Section. 7 November, p.14.

Elsa (2004). Video Games Market Demonstrates over 100 per cent Growth in Six Years. Press Release. The Entertainment and Leisure Software Publishers Association. 1 September 2004.

Federoff, M.A. (2002). Heuristics and Usability Guidelines for the Creation and Evaluation of Fun in Video Games. Unpublished Master's thesis. Indiana University.

Fiore, S.G. (2004). From Designing for Function to Designing for Meaning. Living and Working with Technology. Twelfth European Conference of Cognitive Ergonomics, pp.99-106

Gates, B. (2004). Digital Entertainment Anywhere Event. Speech Transcript. Microsoft (Microsoft.com). October 12.

Green, W.S. and Jordan, P.W. (2002). Pleasure with Products: Beyond Usability. Taylor and Francis.

Harper, R. (ed.) (2003). Inside the Smart Home. Springer.
ISO (1998) ISO 9241. ISO.org.
Jenkins, H. (Undated) Quentin Tarantino's Star Wars?:Digital Cinema, Media Convergence, and Participatory Culture. http://web.mit.edu/21fms/www/faculty/henry3 /starwars.html
Kim, J., Choi, D. and Kim, H. (1999) Towards the Construction of Fun Computer Games: Differences in the views of Developers and Players. Personal Technologies, 3, 1-13.
Lall, B. (2004) Bollywood Embraces Binary Model. Screen International. June 18, p.21.
May, J. and Barnard, P. (1995) Cinematography and Interface Design. In Nordby, K., Helmersen, P.H., Gilmore, D.J. and Arnesen, S.A. (eds.) Human Computer Interaction: Interact '95. Chapman and Hall, pp.26-31.
McCarthy, J. and Wright, P. (2004) Technology as Experience. The MIT Press.
McKendree, J. and Mateer, J.W. (1991). Film Techniques and the Design of Human-Computer Interaction (longer version of paper presented in Proceedings of the 24th Annual Hawaii Conference on System Sciences Vol.II.
New Media for a New Millennium (2004). Movies to Become Unique to Each Viewer. New Media for a New Millennium Press Release (http://www.ist-nm2.org/) 18 October
Norris, A. (2003).Hide and Seek. Guardian Online. 6 February.
Ofcom (2004a). Digital Television Update Q2 2004. Ofcom. September.
Ofcom (2004b). Driving Digital Switchover: A Report to the Secretary of State. Ofcom. 5 April.
Ofcom (2004c). The Communications Market 2004. 11 August 2004.
Ofcom (2004d). The Ofcom Internet and Broadband Update- April. Ofcom.
Olsen, S. (2004). Netflix, Warner Bros. in video-on-demand test. News.com (marketwatch.c-cnet.com.com [sic]). 22 September.
Paquet, D. (2004). Asia looks to a digital future (South Korea). Screen International. June 18, p.21.
Peters, S. (2003). Emotional Context and "Significancies" of Media. In Harper (2003), pp. 79-97.
Peters, S. (2002) Rethinking Networks: Identities and Connectivity in the Mobile Age. Absent Prescence: Localities, Globalities and Method Conference. Helsinki, 10-12 April.
Shackleton, L. (2004). Asia looks to a digital future. Screen International. June 18, p.21.
The Work Foundation (2003). Cost-conscious consumers threaten 3G success. The Work Foundation (theworkfoundation.com). 1 April.
U.K. Film Council (2004). Digital Futures Seminar. 29 January. (Available at http://www.ukfilmcouncil.org.uk/filmindustry/digitalfutures/)
Ward, A. (2004). Will broadband give us television on tap – or is it just a fat pipe dream? (South Korea). Financial Times Creative Business. 30 September 2004, p.8.

THE TRADITIONAL DESIGN PROCESS VERSUS A NEW DESIGN METHODOLOGY

A COMPARATIVE CASE STUDY OF A RAPIDLY DESIGNED INFORMATION APPLIANCE

Steve Gill , Paul Johnson , James Dale, Gareth Loudon, Bethan Hewett and Gareth Barham

Steve Gill, Gareth Loudon, Bethan Hewett and Gareth Barham, National Product Design Centre, Cardiff, U.K. Paul Johnson and James Dale, Nottingham Trent University, U.K.

Abstract: This paper reports on the results of an exercise held at the National Centre for\ Product Design Research (PDR) which is based at the University of Wales Institute, Cardiff (UWIC) in partnership with Nottingham Trent University, and two UK Top Ten design consultancies, Alloy Product Design and PDD. The event was sponsored by the Audi Design Foundation and set out to cover the ground from briefing document to the full design and prototyping of an Information Appliance within 24 hours. The exercise was undertaken by two teams, one based in London at PDD's headquarters and comprising staff from PDD and Nottingham Trent, and another comprising staff from UWIC and Alloy Product Design, based in Cardiff. The latter team had access to an interface development methodology described in the paper while the former did not. This paper will initially concentrate on the activities of the interface design team based in Cardiff, their design strategies and, in particular, their use of the prototyping methodologies developed at UWIC. The paper reports on the structure of the "day", negotiations between the various teams, the consequent concessions and the integration of GUI and hardware aspects of the interface design process. It then examines the results of the Nottingham Trent/PDD team's efforts and compares the approaches and the results. In conclusion it examines the UWIC interface development methodology process's strengths and weaknesses, particularly through comparison with the more traditional design approach undertaken by the other team.

Key words: Interaction, Design, Information Appliance.

1. INTRODUCTION

The concept of designing and prototyping a product in twenty-four hours belongs to Roger Griffiths, a member of the product design programme at *UWIC*. The concept and management are described in Griffiths (2004).

The exercise was undertaken with *Nottingham Trent University*, *Alloy Total Product Design*, *PDD* and *PDR*. When agreement had been reached regarding the goals of the exercise, an approach was made to the *Audi Design Foundation* for funding.

The aim of the exercise was threefold:

1. An educational exercise for product design students to test the creative process, time management and organisational limits of the design process
2. A test of the fast and flexible interface design and prototyping methodologies developed at *UWIC* in response to the challenge laid down by Margolin (1998), Branham (2000) and others
3. A comparison of the approaches of two design teams working on an identical brief within a very short time frame

2. THE PROCESS EMPLOYED AT CARDIFF

The project team was divided into four groups covering Product Design, Interface Design, Computer Aided Design (CAD) work and prototyping/modelmaking.

The twenty-four hours are detailed below. Interface Design and prototyping matters are given precedence because one of the major aims of the exercise was to test *UWIC*'s design and prototyping methodology which seeks to accommodate the types of design thinking advocated by Sharpe and Stenton (2002), Houde and Hill (1997) and Buchenau and Suri (2000). Its underlying philosophies are described in greater detail in Gill (2003).

Audi briefed the team to design a "blue sky" communications device for use by design-aware 18 - 25 year olds interested in extreme outdoor pursuits.

2.1 Mind Mapping

The first priority for the team following the briefing was to rapidly arrive at a series of concepts. Creative Problem Solving technique (CPS), (Van Gundy, 1988)) and De Bono's Six Hats idea evaluation sessions (De Bono (1990)) were run simultaneously to answer this requirement. The teams discussed the way in which any resulting product might work, the social and technological preferences of the target user and how any product would service these. Materials and technology research & interviews with target user group were undertaken by a third sub-team and the results were fed to the two mind mapping groups. Intriguingly the conclusions of both groups were dramatically similar. There were three main conclusions:

1. A "safety product" would not appeal to our intended user.
2. The product should have a competitive element.
3. The product should have a social element.

2.2 Review of creative processes and research data

The two concept generation teams were merged to discuss their conclusions, and more detailed discussions took place informed by reports from the research team regarding new materials plus the information gleaned from the target user group. A large part of the discussion revolved around the functions of the product and the wishes and habits of the user. Some details of control inputs were discussed at this stage but the overall look of the product was not.

2.3 Concept design phase

The chosen concept sketch is shown in *Figure 1*. It includes features agreed by the concept development team including a single-use "panic button" and a wrap-around Light Emitting Polymer (LEP) touch screen that can be removed from the product and attached to the user's body.

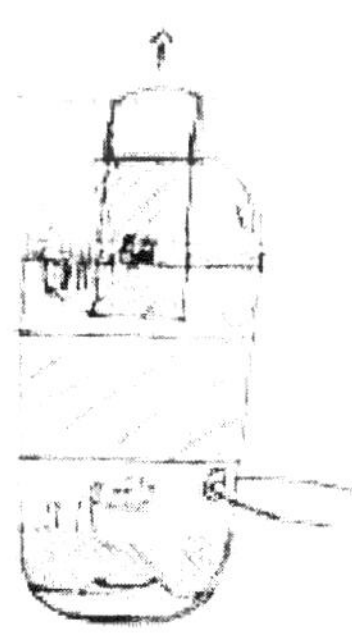

Figure 1. The Cardiff team's chosen concept

2.4 Interface Design

The user interface team consisted of four designers. At 11:00 it separated itself from the other teams and set about conceiving an interface that would deliver to the user the functionality conceived by the design team as a whole. Early on in their discussions it was agreed that the major control input device should be in the form of a rotary dial, because a glance at the position of the dial would make it easy to see the current interface state. This had important implications for the product's design but the interface team felt strongly the importance of this type of control input. A representative from the interface team negotiated changes in the design with the product design team. Further communications were required, to decide on the number of functions and their sequence on the dial. The designs of symbols also required negotiation and discussion between the two teams.

When the basic functions of the product were decided and the negotiations between product and interface design teams satisfactorily completed, the interface design team began detailed design work. The first step in this process was the production of a state transition chart, which effectively became the design specification of the interface prototype. Landay and Myers (2001) report on a similar integration of state transition diagrams in the design development of websites.

The system is described in more detail in Gill (2003) but briefly it involves sketching each state of the product on a *Post-It* note, then numbering each one (see *Figure 2*).

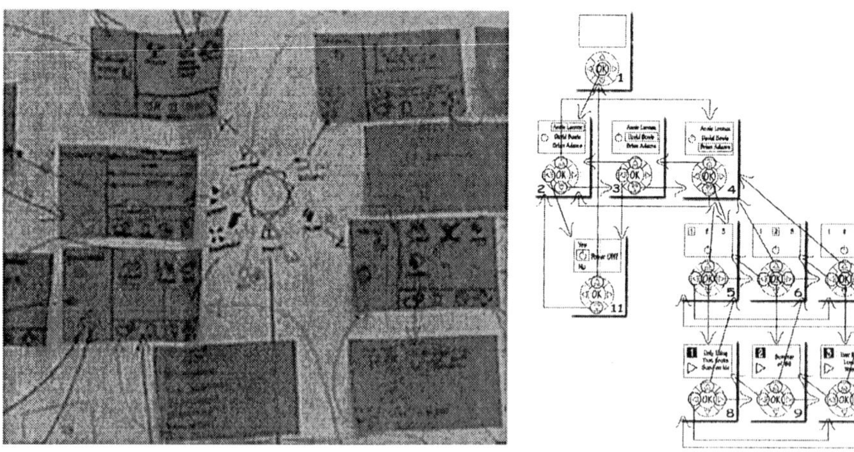

Figure 2. Post-It state transition diagram of the Cardiff Team's interface (left) and a diagram showing how a typical Post-It state transition diagram is laid out

The use of *Post-It* notes allows the designer to change the diagram quickly and easily so that the design is able to evolve as new ideas are inputted to the process.

2.5 GUI Prototyping

When the state transition diagram is complete each numbered *Post-It* is reproduced as a *PowerPoint* slide. Within each of these states there is frequently other work to be completed in the *PowerPoint* prototype that would not have been possible on the state transition diagram, for instance speech to text, text to speech, animations or sound (see *Figure 3*). All states of the GUI are prototyped individually before work commences linking them and detailing the GUI "skeleton".

Using *PowerPoint*'s embedded language, *Visual Basic for Applications (VBA)*, hyperlinks are made linking each of the "states" of the interface together. The control inputs triggering each of these changes in state are effected by a QWERTY keyboard press. The end result of this element of the prototyping work is a PC-based simulation of the GUI where transitions between states are effected by keyboard presses.

2.6 CAD data

With the external surfaces of the product fully detailed the interactive prototype designers and makers could start work. Their job was to create a model which would link up to the GUI prototype allowing potential users to gain a good feel for how the completed product would be in actual use. Sharp (1998) demonstrated that virtual prototyping of touch-screen GUIs could be effective this way. The product in Sharp's study however was a microwave oven with a flat vertical surface. Users interacted with it in the way they would a touch-screen, so a 3D product was not in effect tested. Pering (2002) describes a bespoke method called "the Buck System" used at *Handspring* where an existing product was used to trigger a PC programme. As she points out, however, flexibility remained a serious problem with this method. The *UWIC* method described below is designed to circumvent both these problems.

2.7 Interactive and facsimile models

Once the CAD work for the interactive prototype was completed the files were sent to a CNC machine which began to manufacture the shapes from solid blocks of model board.

In order for the interactive prototype to connect to the PC-based GUI prototype, switches had to be embedded in it. In this case these are simple micro switches attached to a ribbon cable and tested before they are mounted within the model. When wired to a product called an *IE Unit* (Gill 2003) they allow the PC to receive keyboard inputs via the model (see *Figure 3*).

Thus when a user activates a switch in the model, the PC behaves as if it has received a keyboard input so the GUI prototype responds accordingly.

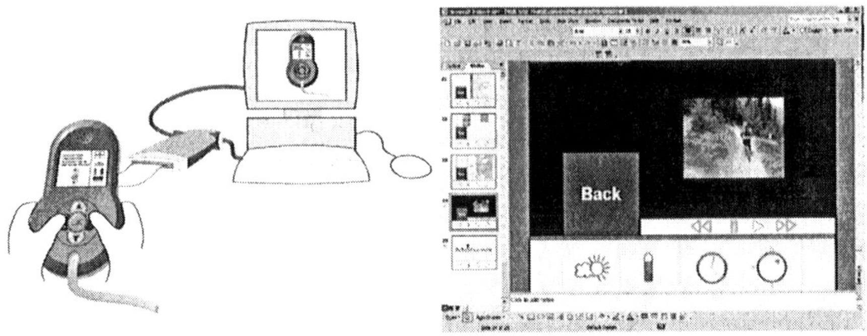

Figure 3. Illustration on the left shows the UWIC IE Unit linking a prototype to a P.C. Illustration on the right shows a single state of the UWIC/Alloy design including graphics, video and sound

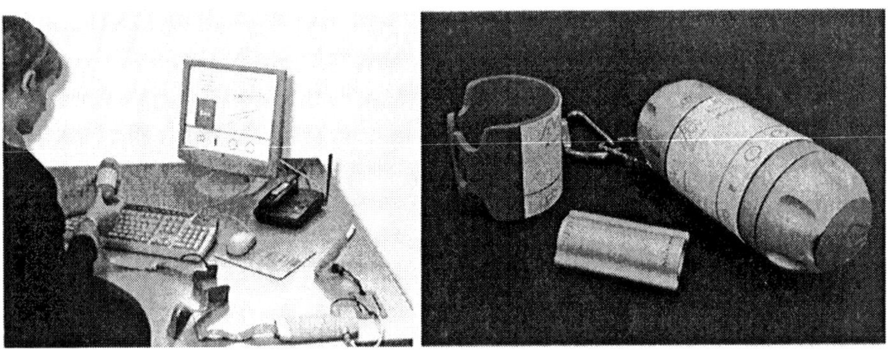

Figure 4. The Mohawk in working prototype and facsimile model forms

2.8 Presentation to user group

The Cardiff team's answer to the brief was *Mohawk* (see *figure 4),* a device for extreme sports enthusiasts of the "*PlayStation* Generation". The device would allow users to record their performance in a "real world" activity (in this case mountain biking), publish it to the internet, keep track of their "tribal" and world rankings, and challenge and meet others from their "tribe". Users could even race "virtually" because the device would have the ability to "ghost" an image of a competitor's experience on top of their own.

Similar approaches to the use of social enabling technology can be found in Vogiazou et al. (2004), Frohlich (2004) and Reid (2003).

3. THE PROCESS EMPLOYED IN LONDON

The methodology and outputs of the two teams varied significantly. Many of *PDD*'s commercial techniques such as "culture hunt", user profile generation, creative brainstorming, brand intelligence, commerciality investigation and focus group reviews were utilised. These were employed to reduce the risks of the product development process and identify product opportunities, understand the needs of consumers and create a brand identity. They are briefly described below.

3.1 Market research via Culture Hunt (thirty minutes)

This activity was carried out by talking to members of the general public to identify the target market for a communications device.

3.2 Profile Generation exercise

Various tools were used to examine the product's user. As a PDD employee stated: "Don't forget we are designing something for a future target market. We are not ourselves the market, and we can't ask those who are now in their pre-teens the relevant questions."

In order to create a profile, a lot of information about the target market was required, e.g. their tastes in music and fashion and their aspirations and interests. Each Product Profile generation team had different age ranges to consider and paperwork with basic headings was distributed to each group to facilitate discussion.

In PDD's estimation the exercise failed to garner enough useful data principally because it did not gather the views of a wide enough cross-section of the target market. Furthermore, what was identified were aspirations rather than needs. They stated: "This represents a very small part of the population. The only learning we can take from this exercise is that we do not know our target market."

3.3 Facilitated Concept Brainstorming Six Hats & Needs Review

One of the advantages of the *Six Hats* method (De Bono (1990)) is that it democratises the design process and prevents dominant personalities from controlling it. Accordingly, the contributions of all members of the team are granted equal status. The process briefly involves participants being asked to think about a given subject from six different viewpoints (thus the reference to six "hats").

The primary aim of the process is to "undo" the way in which the brain is conditioned to think and thereby generate unusual ideas. Each idea generated is recorded on a *Post-It* note. The members of the group review the *Post-It* notes and share their ideas before separating to develop these further on an individual basis. (see *figure 5*).

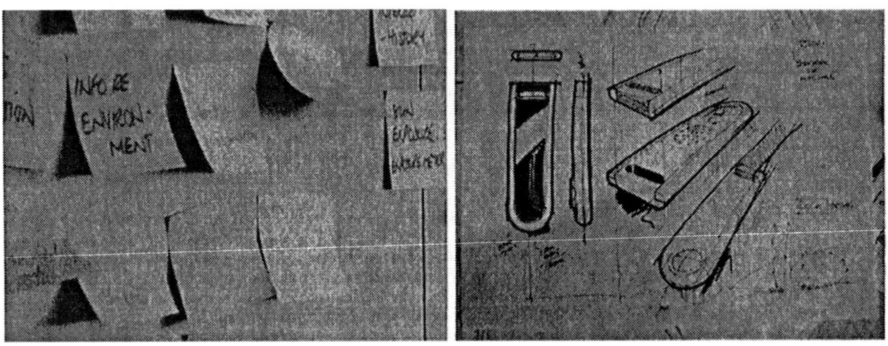

Figure 5. Six Hats & Needs Review and Concept sketch

Members of the team then voted on each others' ideas and rated them by allocating a number of stars. The highest scoring ideas were selected and the authors asked to explain their ideas more fully, resulting in six concepts. Each person on the team was then given two votes and by 12:00 noon the choice had narrowed to three concepts.

3.4 Concept "work up"

Three hours were allocated for three teams to produce a full product design specification, in visual form, of each of the potential ideas. Deliverables included rough 2D visuals and a specification of materials, finish and functionality. In essence these were used as feasibility studies of external form, internal components, interface, interaction and key features.

3.5 Teams present and agree final product

The final stage of the concept selection process involved the three teams making presentations for debate and discussion to each other, the *PDD* technicians, and administrative and marketing staff. The final design was selected by this group as a whole.

3.6 Product detailing

Once a concept had been selected, the design process employed was similar in several ways to that followed in Cardiff. One notable exception to this was the interface design which was tackled with very different methods. Following the selection of the final concept the main design team was split into a series of sub-teams, each of which was set the task of detailing various aspects of the product's design. Interface design issues were dealt with during this phase with each sub-team producing interface concepts as part of the detailing process. These were presented in the form of storyboards which, as well as illustrating the concept's interface proposals, also contextualised its use. Storyboards were presented to the team as a whole and critiques were held which served to "debug" the concepts.

Form and scale were evaluated through the production of foam models and were subjected to design review by a focus group. Once the form and detailing had been agreed upon, the work of producing the finished product was handed to the CAD team for modelling. CAD data was sent for CNC machining. Once this was complete model makers produced a facsimile model of the design that was presented at the project's conclusion. There were some additional processes employed in London however, including packaging design, brand intelligence, storyboarding, commerciality checks and photography. Some of these are described in more detail below.

3.6.1 Brand Intelligence Workshop

The Brand Intelligence Workshop exercise was designed to ensure that the concept's branding created the right associations in the minds of potential users. The exercise was in two phases. The first involved placing 400 cards with images of known brands on them and asking members of the team to choose three cards which possessed the attributes they wished to have associated with the concept (e.g. "excitement" "quality" etc.). Each member laid their choices before the rest of the team and further "filtering" took place. The second exercise involved the team's leader choosing four words appropriate to the concept and writing them on a quadrant (see *Figure 6*). Team members were asked to place selected brand cards in the quadrant that they felt best suited their attributes. Of particular interest were brands that crossed multiple boundaries.

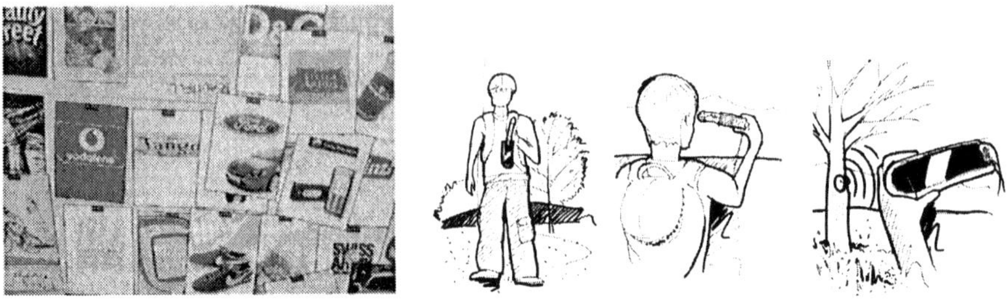

Figure 6. Brand Intelligence sorting grid & storyboarding of the chosen concept

The results of both exercises were then analysed in order to produce an appropriate brand for the chosen concept.

3.6.2 Storyboarding

As well as their role in the interface design development, storyboards were also used to explain and contextualise the way in which the concept worked at the final presentation (see *Figure 6*).

3.7 Touchstone

The London team's answer to the brief was called *Touchstone* (see *Figure 7*). The concept enables people to record a memory or experience onto a server located at points of geographical interest to be accessed by like-minded travellers in the future. Went (2004) describes the concept:

> The final product, Touchstone, was conceived as a digital version of a cairn. Instead of stones to mark and memorialise, Touchstone builds layers of digital experience. Personal memory and data pass between a static digital message-board and a personal handheld unit, allowing images and messages to be left on the Touchstone for others to see and hear. The durable alloy casing, dual display, GPS and short-range radio make Touchstone as useful in town as in the country. The product also includes a keypad interface integrated into an intelligent textile strap, augmented reality graphics, induction recharging, integral imaging and sound recording capability.

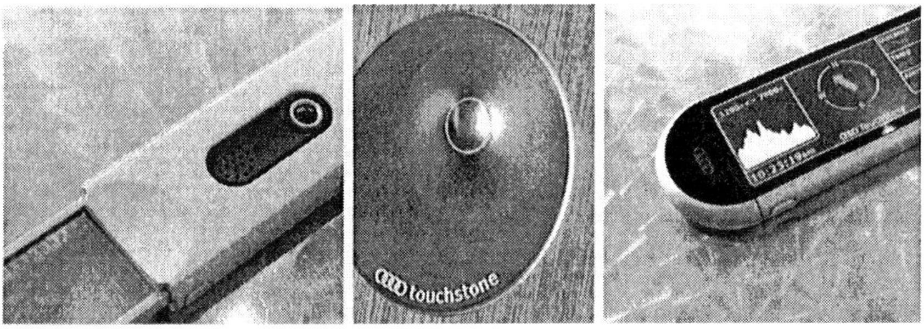

Figure 7. The Touchstone facsimile model

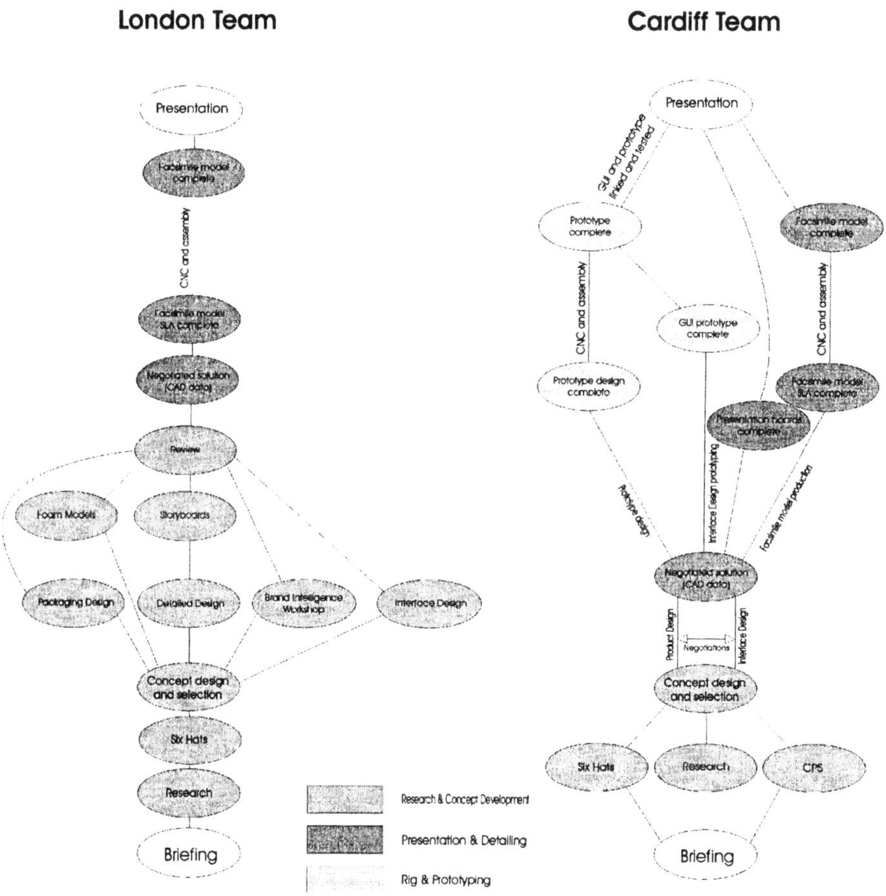

Figure 8. The Cardiff and London Team's Design Processes

4. COMPARISONS OF METHODOLOGY AND RESULTS

The two teams approached the same problem in different ways. There were a number of similarities, both used *Six Hats*, both divided their teams into sub teams to tackle specialist areas of operation and both used regular review meetings to manage and control this process. The main area of difference is that the team in Cardiff were driven from the beginning to deliver a working prototype. The team in London were not and this had important ramifications on the design process and outcome. *Figure 8* illustrates the two approaches.

The London-based team spent a much higher proportion of its time conceptualising and researching. This meant that the product development process was fuller and that considerations such as the end-user, tactility, weight and balance were given greater attention than was the case in Cardiff. Other areas that received more consideration were commercial concerns such as packaging, branding and product specification.

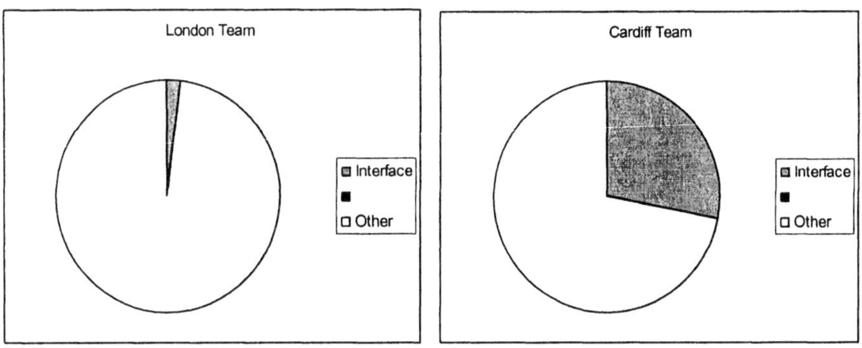

Figure 9. Proportion of team effort on interface design

The team in Cardiff committed to a specific concept and design aesthetic a full 21 hours before the deadline, meaning that this aspect was allocated only 12% of the total project time. The London team did not commit until 12 hours before the deadline meaning that they allocated a full 50% of their project time to conceptualisation, research and branding. In contrast the London team spent approximately 2 hours (8%) using 17% of the total team for that time on the interface design while the Cardiff Team spent a full 20 hours (83% of total time) with a full 34% of the project team committed to this aspect. *Figure 9* shows total team effort on interface design and prototyping by each of the two teams.

The necessity of producing a working prototype forced the Cardiff team to commit to key decisions at an earlier stage so that CAD data could be shared and used on two quite separate operations. From the time the data was "divided", changes to the design could only be made by negotiation between teams, effectively ending any design review opportunities as each team became committed to a particular course of action.

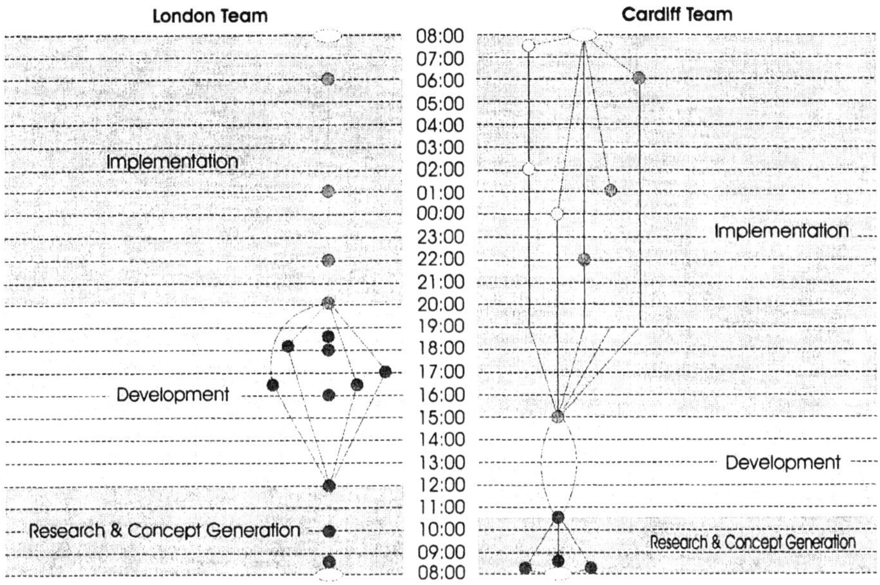

Figure 10. The Cardiff & London Team's workflow to scale against Time

The "trade off" is that the *Mohawk*'s interface design is better understood than that of the *Touchstone*, and this means that the interface and product design were in better equilibrium. However, one never gains without loss, and in terms of a complete, market-targeted package the *Touchstone* is the more complete of the two.

It is important to note that despite the significantly different approaches of the two teams, they arrived at some very similar conclusions. Like the team in Cardiff, the London team concluded that it was important for the product to be a social device that enhanced communication between groups with similar interests. Both teams in fact used similar terminology for this cultural aspect, *Tribe*.

5. CONCLUSIONS

The rapid interactive design and prototyping methodology developed at *UWIC* has been proven to be capable of being implemented within a very short timescale. The Cardiff team showed it was possible to conceive a product, develop its interface and create a three- dimensional prototype with a fully operational GUI inside twenty-four hours. The process influenced the product design process significantly and, while it enhanced some areas, it also forced decisions and resources to be committed earlier than was the case with the London team who considered more closely a number of subtle design matters including tactility, weight, packaging and branding. They were also able to devote more time to researching user needs and wants to ensure fitness for purpose at a conceptual level.

It would be fair to conclude that while the *Mohawk* demonstrated a better balance between product and interface, it lacked some of the simplicity and subtlety of *Touchstone,* arising from a more in-depth analysis of user and design issues.

It could further be extrapolated that the London Team deployed their resources more appropriately than the team in Cardiff. Given the timescale and desired outcome, it was more appropriate to concentrate on the thorough understanding of market placement and the development of the concept to a higher degree of "polish" than to move ahead to interface design issues before these were fully resolved.

In the end analysis, the interface design methodology was proven, but, perhaps, inappropriately applied.

REFERENCES

Branham, R. (2000) Given the radically changing work environment and new worldviews, what kinds of new 'tools' do designers need to survive and successfully deal with tomorrow's design problems? in Proceedings of IDSA 2000 National Education Conference, September 18-20 University of South-western Louisiana, Lafayette, LA, USA

Buchenau, M. and Suri, J.F. (2000) Experience Prototyping, DIS '00, Brooklyn, NY

De Bono, E. (1990) Lateral Thinking – A Textbook of Creativity, Penguin, England

De Bono, E. (1990) Six Thinking Hats, Penguin, England

Frohlich, D. M. (2004) The Audiophoto Desk. The Proceedings of the 2nd International Conference on Appliance Design, May 2004

Gill, S. (2003) Developing Information Appliance Design Tools for Designers. The Proceedings of the 1st Appliance Design Conference, May 2003, HP Labs, Bristol, UK

Griffiths, R. (2004) The 24-hour Product – From Concept to Interactive Model In Less Than a Day, Proceedings of The 2nd International Engineering and Product Design Education Conference September 2 – 3, 2004, Delft, The Netherlands

Houde, S. and Hill, C. (1997) What do Prototypes Prototype? Handbook of Human-Computer Interaction (2nd ed.), M. Helander, T.Ê Landauer, and P. Prabhu (eds): Elsevier Science B. V: Amsterdam, 1997

Landay, J. A. and Myers, B. A. (2001) Sketching Interfaces: Toward More Human Interface Design, IEEE Computer , 34(3), pp 56-64, March 2001

Margolin, V. (1998) Design and the world situation, in Balcioglu, T. (ed.) The Role of Product Design in Post-Industrial Society, METU Faculty of Architecture Press, Kent Institute of Art and Design, pp. 15-34

Pering, C. (2002) Interaction Design Prototyping of Communicator Devices: Towards meeting the hardware – software challenge, Interactions Journal, November and December 2002

Reid, J. (2003) Schminky: The Design of a Café based digital Experience The Proceedings of the 2nd International Conference on Appliance Design, May 2004

Sharp, J. (1998) Interaction Design for electronic products using virtual simulations, PhD Thesis, Brunel University, UK

Sharpe, W.P. and Stenton, S.P. (2002) Information Appliances, chapter from The Human Computer Interaction Handbook, Julie A. Jacko, Andrew Sears (eds), 2002

Van Gundy, A. B. (1988) Techniques of structured problem solving. Van Nostrand Reinhold, New York

Vogiazou, Y., Raijmakers, B., Clayton, B., Eisenstadt, Geelhoed, E., Linney, J., Quick, K., Reid, J., Scott, P., (2004) 'You Got Tagged!': The City as a Playground, The Proceedings of the 2nd International Conference on Appliance Design, May 2004

Went, G. (2004) The Sharpest Tools in the Box, PDD web-based marketing information http://www.pdd.co.uk/pdf/pdd_news_audi-24hrs.pdf

CHAIRS: THE AMBIGUITY OF INVISIBLE OBJECTS

Salvatore Fiore
School of Computing and I.T., University of Wolverhampton, U.K.

Abstract: In this paper, I describe work oriented to better understanding meaning-making within the experiences of blind people. Based on pragmatist aesthetics, the work presented forms part of an extended project towards the development of an interactive chair designed in light of specific ways in which blind people construct meaningful experiences around interactive artefacts. This paper discusses one phase of the research leading up to the construction of the chair that has involved the use of storytelling as an imaginative means of exploring interaction possibilities and experiential quality. The work is aimed at fostering an approach that prioritises designer agency and supports a connection between the different ways of experiencing of a sighted designer and blind users of the artefact. A story is presented and conclusions are drawn for the continuing work.

Key words: design, interactivity, storytelling.

1. INTRODUCTION

Human Computer Interaction (HCI) approaches have traditionally been guided by criteria of measurability of success in supporting human-object interactions, simplification and representation of human and machine behaviours and predictability. People and interaction are modelled in terms of processes and tasks, establish practices that provide for the design of highly efficient, usable and transparent systems. However, recent developments in HCI research concerned with establishing theory and practice that instead take human experiencing as both object and ideation are bringing alternative conceptions of possible approaches to the creation of technological artefacts. Recently, many HCI researchers concerned with experience have turned towards various interdisciplinary paradigms and metaphors, as diverse as Literary Theory (McCarthy & Wright, 2003) (Coyne & Rebelo, 2003), Art (Gaver et al, 2004) and Pragmatist Aesthetics (Petersen et al, 2004) (Fiore, 2003; 2004) (McCarthy and Wright, 2004a; 2004b) for ideation.

The contribution of Pragmatist Aesthetics to such recent advancements in HCI is particularly noteworthy for providing a way of understanding experience in an holistic way. Pragmatism, and particularly the works of John Dewey (Dewey, 1933; 1938; 1958), accounts for the socio-historical, intellectual, physical and emotional dimensions of experience that work together to give an experience its unique and unrepeatable quality. Taking such a standpoint, Petersen *et al* (2004) emphasise equally the socio-cultural nature of artefacts, whereby the aesthetic is "released in dialogue as we experience the world...based on valuable use relations influencing the construction of our everyday life" (p.271), the inseparability of body and mind in meaning-making that contrasts the preoccupations with separating humans into body and mind of dualist perspectives, and instrumentality that links with the notion that for a thing to have value, for it to have the potential to deeply move and provoke a person toward new insight, it must relate to human needs, desires, fears and hopes. In this sense, artefacts achieve meaningfulness through their appropriation in use. Such issues are also emphasised in McCarthy and Wright (2004*b*) and Fiore (2003*a*; 2003*b*; 2004). Such approaches avoid the ad hoc combining of complementary ideologies and instead adopt Pragmatism as an entire alternative perspective on experience, subverting an analytical understanding of computing systems.

However, if this perspective is to foster tangible objects for HCI, more research is needed to examine new modes of practice that support such an holistic understanding of experience. In particular, pragmatist HCI needs to develop methods that account for experience as a phenomenon relevant equally to the creation of artefacts as to their ongoing use and appropriation as meaningful objects into the lives of people. This rests on a notion of the 'designer' not as neutral observer, but as an active, emotional agentive individual in the construction of objects, that will inevitably form the focus for reflective experiences. Dewey's conception aesthetic experience necessitates viewing designers as collaborative agentive subjects, who create through their own reflection and phenomenological being. The object is an expression of the designer's perspective and mirrors their way of seeing the world so that the act of designing becomes itself a means of reflection in and on experience. According to Dewey, we are always seeking to resolve conflict, doubt, and indeterminateness into harmony, employing the emotional-volitional aspects of experiencing as processes of sense-making.

Design might be viewed in this sense as an endeavour that is closer to art than craft, whereby, as Collingwood describes, the goal is not to fabricate a planned emotional effect on others but rather for the artist "...to explore his own emotions; to discover emotions in himself of which he was unaware, and, by permitting the audience to witness the discovery, enable them to make a similar discovery about themselves..." (Collingwood, 1958, p.119). The question then is how HCI might provide methods that support creative subjects in establishing practices to connect to their audience.

2. CREATING INVISIBLE OBJECTS

Despite the pragmatist turn towards experience in HCI, such concerns generally fail to address the particular needs of blind users of technologies, as well as people disabled in other ways. The danger with this is that regardless of the quality and integrity of the research, accessibility may be relegated to the status of a secondary add-on in the design process, involving the functional adaptation of technologies that are really designed with the sense-making processes of sighted people in mind.

From a Pragmatist perspective, it becomes necessary to accept the unique qualities brought to each experience by the subject, their way of perceiving the world as a sensing, conscious being. This raises particular problems for the case of sighted designers aiming to design artefacts that will be meaningful for others who are blind.

Josselson argues for empathy and narrative as "routes to imagining what is real in whole people in their world" (Josselson, 1995, p.29), suggesting that we must first recognise our own role as knower before we can know the real, internalising our observations in the external world so that we might 'tell a story about what we believe we know' (*ibid*). She suggests the possibility to advocate an 'empathic stance' within research which focuses on discovery rather that the confirmation of hypotheses or the pursuit of illusory pre-existing truths; "If we listen well, we will unearth what we did not expect. This becomes the paradigm for discovery" (*ibid*, p. 30). Might not HCI therefore be enriched by methods that support designers in using imagination and empathy to understand something about the way others experience the world differently. To this end, we seek new conceptions of approaches to the creation of objects that suggest meaningful possibilities for interaction for a blind user and help deepen or broaden the objective conditions open to that person to experience in their way.

3. THE RESEARCH

The aim of this research is to further explore these issues with a particular focus on the ambiguity of an interactive object as experienced with or without sight. The main work focuses on the creation of a chair, the qualities of which emerge from the researcher's explorations into the blind perspectives on the world.

As a starting point, various autobiographies by blind people have been examined. Such literary study did not aim to find evidence of specific prescribed hypotheses about blind experience. Instead, the activity was one of allowing relevant issues to be revealed in the reading as I related the texts to my own experiences as a sighted person. A discussion of this work is presented elsewhere (Fiore, Wright & Edwards, 2005).

As a means of reflecting further on the issues emerging through the autobiography readings, I have adopted storytelling as a way of realising imaginative reflections on the experiences of others, in object form. The story presented below was composed as a way of exploration of my own reflections on the life stories embraced as documents of experience. It is not intended to be read as a realistic contrivance aimed at summarising qualities of blind experience into manageable 'key issues' which would be demonstrative of an analytical approach. It is an effort to try to imagine through experiences of blindness as understood dialogically in the reading of the texts of others, opening a phase of active reflection following the reading. The focus as such is on felt experience as opposed to actions or tasks.

The document is essentially the narrative of a sighted person purposefully trying to imagine a world without sight. It connects the two ways of being, so that one might design an object that has the potential to be meaningful to the other.

The story will form the basis for the subsequent stage of the research, which involves the development of a storyboard continuation, aimed at further exploring the object and interaction possibilities. This will further guide the construction of the interactive chair itself. Significantly, the chair itself is not intended as a solution to a problem, but as an object created through exploration that will itself further explore the ambiguities of objects and the particularities of interactions between blind people and the objects that inhabit their world.

4. A STORY ABOUT A CHAIR:

"In the distance, the feint sound of traffic was a buzz in the air, a gentle hum that floated freely over the crystal coated grass, enveloping the open space in sound and yet failing to contain it, continuing its fading stream. The soft breeze carried the clean smells of frozen earth and chestnuts. A cold sun hung white in the cobalt sky, suspended like a distant bulb, its heat a distant memory. Footsteps crunched over the green as the early crowd criss-crossed the space, making towards objects of their desire, ready to bargain or only bemuse. Anna moved cautiously through the emptiness of the open space. To her, the stillness of the air made the field seem as infinite as the sky. Moving towards people, she could hear their conversation already: a man and a woman, talking of some object, reminiscing in shared experiences of a now extinct relic of their childhoods. The woman's voice suddenly seemed to rise, not in volume, but physically lift upward; she had stood up. As other people passed by, Anna caught fragments of their conversation. She turned her attention to her daughter, Katie, who walked steadily and reliably between Anna and her husband Michael, one guiding hand for each. The child led them towards the reminiscing couple, steering her parents skilfully towards items worth exploring at the car boot sale. Anna's guide dog, Lucy, had been given the Sunday off and trotted excitedly on a lead at her left. Michael carried a long white cane to feel his way and rested his other hand lightly on his daughter's shoulder. The family moved silently and trustingly together. They had come early to avoid the chaos around the parked cars and tables.

It took exactly 36 steps to reach the chosen stall. The man and woman had now stopped talking and while he settled curios on a tarpaulin spread over the ground, she returned to her seat to sip hot coffee. As Anna and her family stopped and greeted the woman, the young girl left her mother's hold, returning seconds later with a box carved all over in intricate detail that bewildered even Anna's expert fingers. It smelt like a church; the delicate mingling of petals and aged oak. The family had just moved into a new house. The box would be perfect on the table that stood four steps inside the hallway from the front door. Anna would keep her keys there and the small everyday objects that always seemed to disappear from their carefully rehearsed setting.

Agreeing a price for the box with the woman, Anna turned her attention to the other object they needed for that same place; a chair. The woman indicated the three she had to choose from, but one was exceptional and an immediate choice. The woman recounted the story of the chair; how her grandfather had carved it as a gift for his wife. The cushions she had stitched herself, dressing the naked object, adding herself to its romantic history. Her pride at its artistry was clear and she seemed reluctant to sell the chair. She showed Anna the silkiness of the sculpted wood and invited her to sit. This was an ideal chair for their hallway – comfortable, just the right height for the telephone table and, according to sighted bystanders, very elegant in appearance too. It would serve its purpose well.

The next day, Anna awakes disorientated in the new house. After just a few days, she still struggles to remember the position of everything and the first moments of the day gift her with a new bruise on her leg from a forgotten tea-chest, still packed with belongings and minutes lost gathering together her clothes. It is Monday morning and she is late and frustratingly slow preparing her daughter for school. After her husband and daughter have left, she settles down at the computer to start work. After concentrating on a decidedly one-way conversation with the screen-reader for some hours, Anna decides it is time for shopping. She calls Lucy and goes to the hallway to fetch down the dog's guide harness. And there she stops in her tracks. As she enters the hall, she can hear a vibrating sound ahead of her.

Thinking that her mobile phone has been erroneously switched to silent vibrate, she rushes to the table, searching for the juddering object but finding nothing except the box from the car boot sale. It is not the phone vibrating. Following the noise, Anna moves towards the new chair. As she approaches, the vibration seems to intensify and quicken. Lucy barks once at the convulsing thing in protest and confusion. All sorts of thoughts are going through Anna's mind now. Is it a joke? Is it magic? Is it dangerous...an electrical fault of some kind? What should she do? Why is the chair vibrating? Should she sit? Growing anxious, she eventually sits, and as she does, the vibration ceases. Anna stays motionless. Now she dare not move, unsure of what will happen next. When nothing does, Anna grows restless, wanting to get up and leave. With a rush of confidence, she leaps out, bounding away and stubbing her toe on the table that sits innocently – silently and clumsily – nearby. What is the meaning of this strange chair? Anna picks up the dog harness from the table where she left it. She is oddly nervous to touch even the table: the things around her suddenly seeming so much more unpredictable and unknown. Then she realises that the chair has been silent and still. It does not intrude on any of her four senses as she retreats to the kitchen for a glass of water and time to think.

After some moments, she begins to see some humour in the situation and laughs (a little nervously) out loud to herself. But when she re-enters the hallway, the chair starts up its lunatic dance again. Intrigued now, she begins to play with it, moving closer and retreating; the chair crescendos and subsides. But she is not sure whether she or the chair is really in control. It is predictable now, but she cannot stop it without sitting down. It knows when she is coming and tells her than it knows. She cannot ignore its presence and it dominates the space it occupies like a territorial animal.

Realising that time has been passing during her chair games, Anna exits hurriedly with her dog, the vibrations fading behind her. When she returns, her husband and daughter are already at home. Unsure of what she will find and not prepared to try and explain the strange happenings of the afternoon to them, she enters timidly. The chair is silent. Could it have been a dream? Worse still, could she be developing some unknown hearing disorder in addition to her blindness? No, she reasons; she could feel it too. She moves slowly towards the position of the chair, reaching out to touch it, half expecting it not to be there at all. When she finds it, something is different. Feeling properly, she recognises the fabric of her daughter's coat strewn over the chair. The child has tricked the creature into silence!

Anna hears her daughter emerge skipping and Katie rushes to kiss her mother. Without a word, she pulls off her coat from the chair and almost instantaneously, steps confidently up onto it to reach and hook the coat onto a peg. Anna grasps onto her daughter's waist, worried that the chair will 'wake up' and shake her onto the floor. But it is still. Then, as soon Katie steps down, the chair, seemingly feeling ambushed, shakes madly in the sudden proximity of mother and child. "You have to sit." says Katie matter-of-factly, "It's telling you, you can sit". Anna doesn't quite know how to respond. "Don't be silly. Chairs can't tell you things" and then with suspicion "How do you know its saying that?". For the child, the wisdom to suggest the impossibility of a communicative, even helpful chair, was lacking. She imagined the chair to be friendly and helpful (if a little insistent). "Daddy said if they had one like that at work, then he wouldn't need his cane to find his desk any more. The chair could show him the way to it". This bewildered Anna further; could her husband also believe this fantasy of a 'living' chair?

Over the next few days, Anna's feelings of distrust and uncertainty towards the chair began to dissipate. She stopped wondering about the secret mechanism causing the chair to vibrate and accepted its presence. She even began to enjoy it. She had come to enjoy sitting on the chair when 'invited' and would sit talking on the phone there. Spending time with the chair, she had even come to learn new possibilities with it. Most interestingly, she discovered that the chair would indicate the presence of another person to her: there she would be sitting when a warmth would radiate from the seat, accompanied by an almost imperceptible tremble. As the person drew nearer, she found herself feeling less surprised by their sudden emergence out of nowhere and more prepared to try to recognise them before they spoke to her. Sitting in that chair, the element of surprise that her blindness forced upon her in everyday encounters, was diminished. The space around her seemed mapped by the radiant sense of the chair.

For some weeks, Katie's friends would play around the chair, teasing it, trying to creep up on it unnoticed and delighting in it's sudden convulsions when they were 'discovered'. But they soon grew tired of the predictable object and moved on to new and different toys. Anna, however, began to have new ideas about the chair. She could not forget her husband's comments about using the chair at work and one day suggested that he ask his brother to take it there for him to use.

The first few days that Michael had the chair at his desk, his colleagues found the whole thing amusing. They jostled to try and see how it worked and invented jokes about this new lively thing. Before long however, they began to feel quite differently about it. When they approached Michael, he now knew that they were coming without the need to speak. It was 'seeing' their approach for him. The biggest changes were for Michael though. It was a sense of control and freedom. He came to know it's signals subtly. Sitting there, he could feel the hesitancy in the approach of an indecisive colleague or the speed of an eager friend, coming over to talk. It was something like the signals Anna had told him she felt from Lucy the guide dog; she knew when the animal was reluctant or confident from the tiny changes in the position of the harness and the speed of her gait. In the sometimes noisy office, the chair helped Michael feel more at home in his surrounding space. It gave him back some unexplainable quality of the sense he had lost all those years before. The difference was small, but for Michael, it was so very important. He thanked Anna for this gift."

5. CONCLUSIONS

Writing the story was a means of reflection that brought some issues to the fore, which might otherwise have been lost through an analytical approach. The exercise was one of exploration for the designer to work imaginatively to consider the implications and interaction possibilities of such an object for a blind person. The focus was not on setting specifications or extracting requirements from intended 'users' but on exploring the nature of experience and the place of certain artefacts in the shaping of aesthetic experiences of blind people in everyday situations.

REFERENCES

Brace, M. (1980) Where there's a will. Souvenir Press

Coyne, R. (1999) Technoromanticism: Digital Narrative, Holism, and the Romance of the Real. MIT Press.

Coyne, R. and Rebelo, P. (2003) Resisting the Seamless Interface. HCI, Arts & Humanities workshop, University of York

Dewey, J. (1958) Art as Experience. Capricorn.

Dewey, J. (1938) Experience and Education. Macmillan.

Dewey, J. (1933) How we think. D.C. Heath and Company.

Dourish, P. (2001) Where the Action is: The Foundations of Embodied Interaction. MIT Press.

Dunne, A. & Raby, F. (2001) Design Noir: The Secret Life of Electronic Objects. Birkhäuser.

Fiore, S. Wright, P. & Edwards, A. (2005) Agency, interaction and disability: Making sense through autobiographical accounts. Proc. Include 2005.

Fiore, S. (2004) From Designing for Function to Designing for Meaning. Proc. ECCE-12.

Fiore, S. (2003a) Supporting design for Aesthetic Experience. HCI, Arts & Humanities workshop, University of York.

Gaver, W., Boucher, A., Pennington, S. and Walker, B. (2004) Evaluating Technologies for Ludic Engagement. Available at http://www.cs.cornell.edu/people/sengers/ReflectiveHCI /proceedings.html.

Harrison, S. and Dourish, P. (1996) Re-Place-ing Space: The Roles of Place and Space in Collaborative Systems. In Proc. CSCW'96, pp.67-76. ACM.

Hocken, S. (1988) Emma and I. Time Warner.

Hull, J. M. (1991) Touching the Rock: An experience of blindness. Arrow.

Josselson, R. (1995) Imagining the Real: Empathy, Narrative, and the Dialogic Self. In Josselson, R. and Lieblich, A. (eds.) Interpreting Experience: The Narrative Study of Lives (vol. 3). Sage Publications. Keller, H. (1996) The Story of My Life. Dover Publications.

Kuusisto, S. (1998) Planet of the Blind. Faber and Faber.

Magee, B. & Milligan, M. (1995) On Blindness: An exchange of letters. Oxford Univ. Press.

McCarthy, and Wright, P. C. (2004a) Technology as Experience, MIT Press.

McCarthy, J and Wright, P. (2004b) Putting 'Felt-life' at the centre of HCI. Proc. ECCE-12.

Petersen, M. G., Iversen, O. S., Krogh, P. G. & Ludvigsen, M. (2004) Aesthetic Interaction – A Pragmatist's Aesthetics of Interactive Systems. In Proc. DIS2004.

TELEPHONE ACCESS FOR DEAF PEOPLE

Alistair D N Edwards
Department of Computer Science, University of York, U.K.

Abstract: The telephone is the most important piece of personal communication technology in the home. It is a technology that is accessible to nearly all people – except those with hearing impairments. Speech recognition and synthesis technology might be used to make telephone communication between a deaf and a hearing person possible. The necessary speaker-independent speech recognition technology is not currently available, but this paper reports a study in which such technology was simulated in order to test the feasibility of such communication. The results demonstrate that such a system would be highly desirable, but it will not be feasible until speech recognition rates are greatly improved.

Key words: Accessibility, deafness, telephone use.

1. INTRODUCTION

The telephone is probably the most used piece of every-day home communication technology. Unlike a lot of technology, it is accessible to almost all users – with the obvious exception of those with severe hearing impairments. The very usefulness and ubiquity of telephone use makes the handicapping effect of not having access that much greater. Improved technology could make telephones more accessible to people with hearing impairments, but there remains a question as to how useful such technology would be. This paper reports a small experiment in which automatic speech-to-text and text-to-speech technology was simulated in order to assess the utility of such a system.

2. THE PROBLEM

Of course there have been vast changes in telephone technology and usage in recent years. Greatest of these has been the introduction of mobile telephones. Their mobility is an important characteristic, but an additional feature that they introduced (almost as a side-effect – see Ocock, 2002) is that of text messaging (Short Messaging System, or SMS). This is probably the greatest innovation in telephony for deaf people. It adds a non-auditory – and thus accessible – channel to the telephone. Most importantly it is a feature available in *all* mobile phones[13] as used by the vast majority of the population. In other words, every mobile phone user has the facility to 'telephone' deaf people – without having to acquire any additional equipment. This has been a boon to deaf people.

SMS messaging is not the same as telephone conversations, in that the communication is *asynchronous*. That is to say that it is not interactive. One person composes a message and sends it to their friend. The message is received on the recipient's phone some time later (and there can be significant delays, because SMS is given a lower priority in the network than voice transmissions). The recipient will read the message at some time – but again there may be some delay. The recipient then decides whether to reply, and if so how (i.e. he or she may decide not to use SMS to reply, but to phone or go to see the person). This is very different from a telephone conversation, which takes place *synchronously*. That is to say that – as long as the recipient chooses to answer – then they are obliged to take part in a two-way conversation. This level of interactivity has a number of useful features: understanding and agreement can be negotiated quickly, for instance. (Reasons why people chose the asynchronous option of SMS over voice – and vice-versa – are discussed in Ocock, 2002).

Even though the use of mobile phones has affected the take up of landline telephones, the landline is likely to be around for some time yet. Thus, there still is an important role for voice telephony – for deaf and hearing people. For deaf users, the current solution to the inaccessibility of the voice telephone is a text-based alternative, based on the conventional telephone network. A *minicom*[14] is a small terminal with a keyboard and a one-line text screen. Two minicom owners can connect their devices over the telephone network, so that the words typed by one can be read off the

[13] This is not strictly true in that SMS was not a feature available in older generation systems, notably systems which were prevalent in the United States for some years.

[14] Minicoms are also known as textphones, or – in the USA – as Telephone Devices for the Deaf or TDDs.

screen of the other. This enables two deaf people to communicate. (Figure 1).

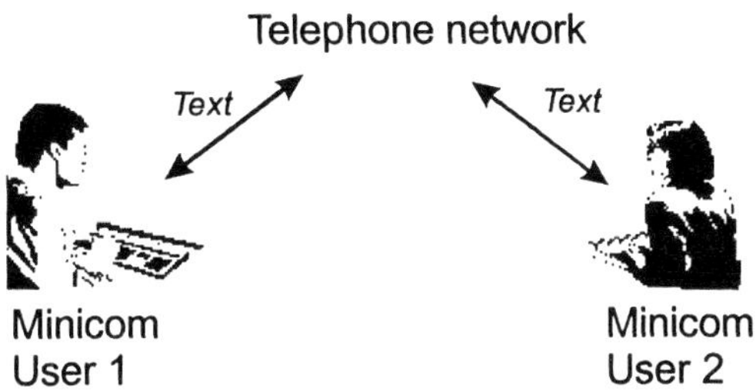

Figure 1. Two minicom users communicating. Both use the minicom keyboard and screen, connected through the conventional telephone network.

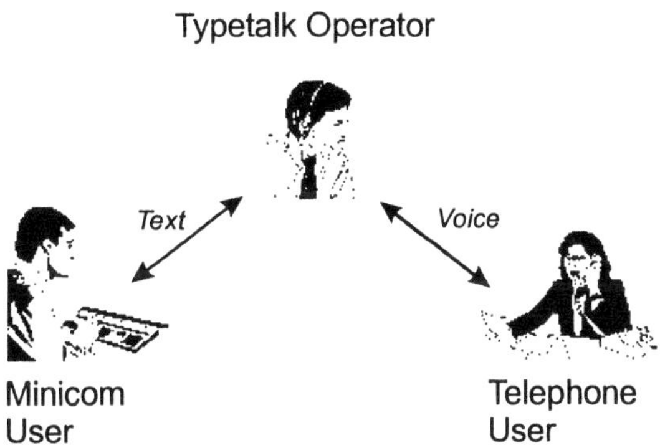

Figure 2. A Minicom User (deaf) communicating with a (hearing) Telephone User. This takes place via a human operator. He must type the spoken messages to the minicom user and read his input to the telephone user.

In principle the same system will work for a deaf person communicating with a hearing person – but few hearing people own minicoms. The solution in this case is to use a relay service. In such a system, two people (one with a minicom and the other with a phone) communicate via a intermediate human operator, as depicted in Figure 2. The words typed by the Minicom User are read out by the operator to the (hearing) Telephone User. She speaks her reply and that is typed back to the Minicom User by the operator. In the UK the *Typetalk* relay service is provided by the Royal National Institute for Deaf People (RNID)[15].

[15] http://www.rnid-typetalk.org.uk/

Of course, there are a number of disadvantages to this solution, not the least that it is impossible to have a confidential conversation. It would be good, therefore, if the human intermediate operator could be replaced by technology. Speech recognition would be used to translate the spoken words into text, displayed on the minicom screen, and a speech synthesizer could pronounce the words that the deaf person typed to the hearing person. (Figure 3).

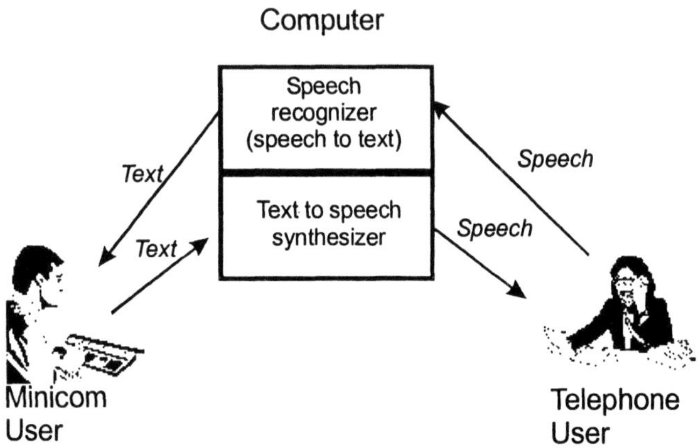

Figure 3. The ideal solution. The human operator is replaced by technology. A speech synthesizer will read out the text that the Minicom User has typed and a speech recognizer will convert the Telephone User's speech into text on the minicom screen.

While this solution has a lot of attractions, it is not currently feasible technically. In particular, the speech recognition engine for such a system would have to be capable of recognizing the speech of *any* person calling the system. In other words, the speech recognition must be *speaker-independent*. Current speech recognition software is quoted as having recognition rates of the order of 85-95% accuracy – but that applies only when the system has been trained to the individual's voice and manner of speaking. Training takes around 5-10 minutes, which would clearly be an impractical prelude to every phone call.

However, being optimistic, it must only be a matter of time before reliable speaker-independent speech recognition is achieved. The idea behind the study presented herein was to simulate that situation and thereby to test the likely effectiveness of the kind of facility depicted in Figure 3. It is often difficult to anticipate the effects – good and bad – of the introduction of new technology. Questions often remain unanswered, such as: Will users want (and pay for) the technology? How 'good' does the technology have to be in order to be successful? Will there be consequences of its adoption that are not anticipated? This is an example of a study which tries to avoid such uncertainty, by simulating a technology which is not currently available – in this case speaker-independent speech recognition[16].

3. SPEECH RECOGNITION

There have been great advances in recent years in speech recognition technology. Early systems had difficulty *segmenting* words, so that speakers had to insert an unnatural pause between each word spoken. Continuous speech recognition – which obviates the need for such pauses – has been a great advance. As mentioned above, recognition rates for such systems are quoted to be around 85-95%. At first glance this may seem to be technically impressive and sufficiently high to be useful. However, in practice such rates are too low to be useful in many applications; the time spent correcting errors and their effects outweighs the advantage of using speech recognition in the first place. On the other hand, it might be argued that in applications for which there is no alternative, such rates are acceptable. In other words, being able to recognize 85% of the words is better than 0%!

Another question is how and whether these rates are attainable. As will be shown below, in this experiment the rates were somewhat lower, much to the detriment of the results.

4. METHOD

The idea behind this experiment was to simulate the set-up depicted in Figure 3 and to carry out a number of conversations between a deaf minicom user and a hearing telephone user in order to find out whether the basic idea seemed viable.

The problem of speaker dependence was avoided by using a single person to act as the telephone user. In other words, before the experiment with both participants, the Telephone User had trained the (speaker-dependent) speech recognition package (*Dragon Point & Speak*, English Version 3.5[17]) to his voice. In principle, the speech output to the hearing Telephone User could have been generated using a speech synthesizer.

[16] It might be said that this study continues the tradition of Dye *et al.* (1989). At a time when any form of automatic speech recognition was not available, they simulated such a system using a skilled human dictation-taker.

[17] http://www.scansoft.com/

However, in practice it proved impossible to use a real speech synthesizer, and that had to be simulated by a person too (as explained below).

The system was simulated using two PC computers which were linked via a TCP/IP network and running *Lan Talk Pro* chat software[18]. The plan had been to use *Jaws*[19] screen reading software[20] to read the text out to the Telephone User. However, there was a technical problem in that whenever a new line of text appeared on the screen, that line alone was not read, but rather the entire contents of the window. In other words, on every exchange the user would be forced to listen again to parts of the conversation they had already heard. It was therefore decided to simulate the speech synthesizer using a human reader. This had the added benefit that it effectively simulated a very high quality synthesizer. The computers were situated in separate rooms in the Department of Computer Science. The 'speech synthesizer' sat in the same room as the Telephone User who had no sight of the screen.

Both participants were given details of a number of role-play scenarios which they were to act out using the system:

- Scenario 1 – Arranging a meeting
- Scenario 2 – Purchasing cinema tickets
- Scenario 3 – Getting to know each other better

After the experiments, each of the participants was given a questionnaire to assess their reaction to the system.

5. PARTICIPANTS

The experiment was carried out twice. The Telephone User in both cases was the same. He had normal hearing and trained the recognition software to his speech. In Experiment 1 the Minicom User (referred to below as MU1) was an undergraduate student who was a Deaf sign-language user (British Sign Language, BSL) who had extensive experience of using minicoms (both in direct communication and through Typetalk). In Experiment 2 the Minicom User (MU2) was a professional sign language interpreter, who had been profoundly deaf, but now had a good level of hearing, thanks to the fitting of a cochlear implant (Tyler, 1993). Since she was born profoundly

18 http://www.thaicybersoft.com/

19 http://www.freedomscientific.com/index.html/

20 A screen reader is a piece of software for blind users which presents text on a computer screen in synthetic speech. (Edwards, A. D. N. (1991). *Speech Synthesis: Technology for Disabled People*. London, Paul Chapman.)

deaf, she was brought up using BSL and taught to lip read. Despite having an implant, she still takes advantage of her lip-reading skill when conversing face-to-face with others. For telephone conversations she has to use a stereo speaker/amplifier attached to her telephone.

(1) Telephone User's speech	*(2) Text displayed on Minicom screens*
	good afternoon, <Name> it is <Name> here
Hi, I've read your CV and I'm interested in interviewing you	*yesterday and then to higher eyes ledger CV hands I mentioned interviewing a further drop*
	i'm not sure if i understand you
I have read your CV and I would like to interview you	*woman's I have Rachel CV and I would like to interview you*
Can you hear me?	*KD here in*
	right, let me have a look at my diary
I'm free on Wednesday the 29th August for the entire day. Would that suit you?	*I'm free on Wednesday the 29th August for the entire day without seeking*
	i'm also free on that day
Okay, does ten o'clock seem a reasonable time?	*okay does ten o'clock seem reasonable time*
	what time would be appropriate for you?
I think 10 in the morning would be good for me	*of 10 in the morning with a good for main*
	that's fine with me
Our offices are in the York Science Park	*okay how offices are in the York assigned sparkling.*
If you ask at the reception they should be able to direct you to where the interview will be taking place.	*if you Askew reception they should be able to direct human to wear an interviewer be taking place*
	right, thank you
Okay, I shall see you on Wednesday 29th August at 10 o'clock in the Science Park.	*okay I shall see you on Wednesday 29th August 10 o'clock in the science park*
	ok then, thanks. bye bye
Goodbye.	*Goodbye*

Figure 4. Transcript of one of the conversations. Column 1 shows what the Telephone User said. Columns 2 shows how the conversation proceeded – as displayed on the screens – including misinterpretations. The Minicom User's (MU1) input – as spoken by the 'speech synthesizer' – is shown in plain typeface (and retains his original informal style of writing) while the translated Telephone User's speech is in italics.

6. RESULTS

The most significant result was that the speech recognition rate of the software was poor – despite the fact that it had been trained to the individual speaker. Recognition rates over the whole experiment, averaged 66%. Figure 4 shows the dialogue for one of the scenarios. It is apparent that there were many errors in the speech recognition. Some of them caused real misunderstanding, but for the most part the Minicom User was able to spot sufficient keywords to understand the intended meaning. It was noted that simple transcription errors were not the only problem. More subtle aspects of speech, such as prosody are also lost. Thus, for instance, 'Sorry?', intended as a request for clarification, appears as an apology when transcribed as 'Sorry'. Prosody is generally used as a means of signalling that the speaker has finished and is expecting a response. This was also lost, causing turn-taking errors and prompting the suggestion that there should be some kind of explicit signal that the speaker has finished the current utterance.

In the post-test questionnaires all participants agreed that the exchanges were time-consuming as a result of the frequent misinterpretations. However, according to the minicom users, the delays were insignificant in comparison to those experienced with Typetalk. For the Telephone User the delays were 'disconcerting' and 'frustrating'. He suggested that in a real phone call he would have discontinued the conversations in the belief that he was the victim of a crank call. His suggestion to eliminate such a problem was to notify the hearing person that a deaf person was on the line by using a 'canned' warning preceding the conversation.

MU1 was overwhelmingly positive about the system and appreciated that communication was directly between him and the Telephone User, that there was no third person involved in the conversation and total confidentiality could be assured. Again, in comparison to Typetalk, the system's speed was impressive – indeed, MU1 said, 'Speech recognition is like instant response. With Typetalk you wait for quite a while when the operator types the response'.

All participants agreed that they found the system to be a good idea and would use it in preference to Typetalk, subject to a number of conditions. All concurred that the delays meant the current system was appropriate for certain control situations, but quite unfeasible for conducting conversations and anything unstructured.

7. CONCLUSIONS AND RECOMMENDATIONS

It might be said that this experiment is flawed in that the component that turns out to have been the weakest link – the speech recognition – was precisely the part that was being simulated. However, the results do have important validity because they clearly highlight the fact that any (speaker-independent) speech recognition to be used in this application must be *very* reliable. That is to say that it will be unwise to attempt to introduce the technology at a point when it is merely adequate – and generate a negative backlash; better to wait until it is highly accurate and so acceptable. (Khine, 2001, lists specific recommendations as to the kind of improvements that would be most beneficial).

It might be argued that the experiment was too harsh, that too much was expected of the speech recognition. For instance, the results might have been better had more care been taken with the use of the microphone and training of the Telephone User. However, this would be artificial in that true conversations would be taking place using telephone equipment with untrained users.

The most important conclusion is that the proposed system does have great potential. The potential for truly confidential conversations cannot be under-estimated.

ACKNOWLEDGEMENTS

This paper is based on the work of Beatrice Khine, undertaken as the project component of her MSc (Khine, 2001). The invaluable assistance of the (anonymous) experiment participants is gratefully acknowledged.

REFERENCES

Dye, R., Arnott, J. L., Newell, A. F., Carter, K. E. P. and Cruickshank, G. (1989). *Assessing the potential of future automatic speech recognition technology in text composition applications.* Proceedings of Simulation in the Development of User Interfaces, Brighton.

Edwards, A. D. N. (1991). *Speech Synthesis: Technology for Disabled People.* London, Paul Chapman.

Khine, B. H. (2001). *An Evaluation of Human Interaction Factors in Deaf Hearing Telephony.* MSc(IP) Project: University of York (http://www.cs.york.ac.uk/%7Ealistair/projects/reports/pdf/khine.zip).

Ocock, M. (2002). *Why are text messages so popular?* York: University of York, Department of Computer Science Third-year Project Report.

Tyler, R. S., (ed.) (1993). *Cochlear Implants: Audiological Foundations* San Diego: Singular.

COMPUTER GAMES AS HOMEWORK
How to delight and instruct

C. Dormann, J-P. Fiset, S. Caquard, B. Woods, A. Hadziomerovic, E. Whitworth, A. Hayes, R. Biddle
Carleton University, Ottawa, Canada

Abstract: We are interested in exploring how entertainment in games can be combined with educational goals to make a compelling experience. In this paper we present our design study for the development of a mod game, Antarctic NWN. We first present the background for the game, objectives, and then discuss the gameplay of Antarctic NWN. We then explore issues that influence the design of a gripping game. One important issue is the relation between reality, simulation and game word. Then we focus on enhancing emotional involvement. Emotion is especially relevant to role-play game as it draws players into the story, and supports aesthetic understanding. We also look more specifically at the role of humour in this context. Humour enhances learning as well as providing a more pleasurable experience. In our quest to understand how games can both delight and instruct, we review the environment in which our game might be played, within the classroom *or as family entertainment* and describe different scenarios of use.

Key words: Computer games, education, entertainment, emotions.

1. INTRODUCTION

One of the main uses of computers in the home is for computer games. Computer games are very popular: in their study, Phillips et al. [1] showed that 75% of children played video games, alone or with friends. While there are games with an education focus, the main reason reported for playing games is enjoyment: games are engaging and fun. In this paper we describe the initial stages of our study of modifying a successful entertainment computer game to play an educational role.

The amount of time that children spend playing computer games at home suggests closer consideration of the role of games. Despite early concern about computer games and the negative effects like violence, addiction or poor school performance, studies have also shown a positive side [2]. Playing computer games can enhance visual processing skills, including visual attention, and the ability to manipulate objects or mental images through space [3]. Players discover strategies for overcoming obstacles, and construct understanding of complex systems through experimentation in natural and unobtrusive ways.

There are countless games where educational objectives are in the foreground, and learning is the explicit goal of the game. There are also games where the educational objectives are more indirect. For example, "Where in the world is Carmen Sandiego" taught simple geography whilst users were playing a mystery game.

We are interested in exploring how games intended primarily for entertainment might also accommodate education. Our current project is to take an existing and popular entertainment game, Neverwinter Nights, and while retaining the entertainment focus, re-purpose it with a new goal of teaching about environmental issues. Our overall objective is to determine how much of the compelling nature of computer games can be combined with educational goals. If this is possible, it could change the role of computer games at home. We thus explore developing a game such as our as a family game. In this paper we focus on basic design issues, reporting on how well we were able to accommodate our educational agenda within the game, and discussing the issues that arise.

The paper is structured as follows. In the next section we outline the background of our project, including our pedagogical strategy, our educational objectives, and the role-playing game we selected to re-purpose. In section 3 we discuss the key elements of our game: environment, characters and the gameplay. In section 4 we examine design issues that are particularly relevant to the design of compelling games: the interplay of fantasy and simulation, and the role of emotional involvement and humour. In section 5 we discuss contextual issues relating to educational setting as well as home and family entertainment. In section 6 we review our experience and present our conclusions.

2. BACKGROUND FOR THE GAME

In this section, we situate the game development within the cybercartography project and discuss the idea and the perspective driving the game design.

2.1 Game Background

Our game study is part of a larger project in cybercartography, a new approach to cartography that "aims at transforming all kinds of cultural, socio-economic and environmental information into interactive representation, allowing users to explore spatial patterns and relationships into new ways" [4]. One goal of this project is a cybercartographic atlas of Antarctica. Some of the content of the atlas is for secondary school students who take part in the Students on Ice (SOI) programme, which features a field trip in the Antarctic. The learning objectives for our game were chosen with these students in mind, and more generally to draw teenagers into the world of Antarctica and environmental issues.

Antarctica is a powerful venue from which to study global warming and its consequences. For instance, ice cores collected from drilling deep into Antarctic ice provide scientists with a wealth of information about past climatic conditions extending over half a million years which help them to understand more recent climate change. The impact of recent global warning has been demonstrated on the Antarctica peninsula [5]. One of the most controversial points remains the potential impact of global warming on sea level. For some this impact is negligible, while for others the sea level could rise up to 80 meters [6]. Beyond this kind of controversy, the risks imposed by global warming on specialised species such penguins are widely acknowledged. These species are highly sensitive to environmental changes and their survival is at stake.

Based on these different elements we highlighted three major pedagogical goals for the game: (1) help the students to learn about the environment of Antarctica (e.g. geography, climate, wild life), (2) expose them to some issues related to global warming (e.g. facts and developing a critical perspective) and (3) expose them to the scientific processes used to study global warming in Antarctica (e.g. ice core analysis). Based on these objectives we are looking at ways of motivating students by making learning engaging. An important goal of this game is to stimulate interest so as students pursue their quest for learning, for example by consulting the Atlas of Antarctica and changing their world view through the exploration of environmental issues.

2.2 Context of Use: Game as Homework

Teenagers are more likely to use information technology at home than at school and form a more positive attitude towards home computing [7]. Moreover, utilising computers for learning, as it is imposed upon students in traditional class setting, is simply not a fun activity [8]. We are not trying to produce a conventional educational software such as computer-aided learning, but to have high school students play a game to stimulate their interest in environmental issues. We initially envisage that the game would be played at home as homework. Despite being set as homework, it is hoped that the home environment would draw the game toward a more leisurely activity, offsetting the negative affect of school work. Moreover having students play at home has other advantages. The short traditional schedule of class is not compatible with gameplay; at home students can explore the game universe freely, and learn at their own speed. In section 5 we explore in more details different context of use including developing Antartic NWN as digital family entertainment.

2.3 Rationale for the Game

Our aim is to combine the learning objectives within the context of an existing successful game. The approach we take is discovery learning, where learners interact with their environment by exploring and manipulating objects, wrestling with questions and controversies and performing experiments [9]. Discovery learning is particularly relevant to our geographical context, but is also at the heart of learning science through aesthetic experience. According to Girod and co. [10], an aesthetic experience further brings students into the world of science through interactions and explorations and the excitement of meeting cognitive challenges. Rich opportunities to explore, engage, and reflect must thus be provided. An aesthetic experience is also emotional and dramatic, and draws in participants.

There are games that support this approach by providing a platform to develop aesthetic understanding with a compelling nature. Such games stimulate exploration through interaction, for example with characters, plots, action, and fulfilment (e.g. winning) that engage players.

We reviewed many successful games, and decided to work with games that offer controlled situations within a simulated world where cybercartography and environmental issues can be explored. Our aim is to create a powerful drama in which students will get immersed, that will challenge them to think through issues to reach their own conclusion.

2.4 Game Platform

For our project we selected the game "Neverwinter Nights" by Canadian games developer Bioware [11]. Neverwinter Nights (NWN) is widely distributed internationally, available on a range of operating systems, and is both popularly and critically successful. NWN is a role-playing game based on the pen-and-paper game "Dungeons and Dragons". The player takes the role of a character who must explore the simulated world, perform tasks, and fulfil quests, with the collaboration or hindrance of other characters within the game. Like many such games, NWN allows users to extend the basic game by creating new regions and new characters. This feature appeared to offer the opportunity for us to retain the popular character and engaging nature of the game, while allowing us to extend the game to support our learning objectives.

3. ANTARCTIC NWN DESIGN

Our intent is to explore retaining the nature of NWN while using this nature to further our educational goals. In particular, our immediate work was to explore whether this was possible at all: was the nature of NWN sufficiently adaptable? NWN is a role-playing game set on a medieval world. We decided to try and build a version of the game where the player takes on the role of a student doing a field-trip in Antarctica (i.e. like the SOI students). We studied the structure of gameplay in NWN, where one of the primary supported concepts is that of quest fulfilment, and decided to use quests to support our educational goals. Our Antarctic NWN would follow the basic structure of a mystery-adventure game. Players are given a basic enigma "the fate of a small island's inhabitant" which is to be solved through a series of quests. The setting for the quests is an Antarctic research station and surrounding territory. Players meet and interact with both human and animal denizens, and are taken to fictitious spaces, as well as areas based on real world locations. To keep and build on the spirit of NWN gameplay, players are given rewards or "magic tokens" which enable them to gain status or solve more difficult quests.

3.1 The Game Environment

Our model world is centred around an Antarctic scientific research station "Mac Town", which is modelled after the real American base, MacMurdo Station. It features buildings such as dorms, pubs, science labs, etc. We have modelled the type of existing buildings and the layout of the station, but not quite its appearance (see figure 1a for NWN MacTown). We are retaining the medieval look of the buildings until we have more experience with the game and are ready to invest in 3D modelling necessary for more realistic buildings.

The game also contains frozen landscapes, including underwater representations: these are not really accurate representations of Antarctic landscapes, but close enough to postpone more faithful 3D modelling. The larger structure of the territory does not follow Antarctica faithfully either, and it is unlikely it will: the terrain modelling in NWN is too primitive, and the true scale of Antarctica would be difficult to accommodate in a game. Simulating a realistic environment is not necessary the design solution as discussed in section 4.1.

The game is populated with characters such as students, scientists, support personnel and appropriate fauna (e.g. penguins); in some cases the appearance of the characters is suitable, but we may eventually need to invest time in 3D design to model Antarctic clothing. The actions of non-player characters are governed by scripts, and the NWN system allows us to make scripts so that these characters can behave like students, scientists, and so on, and also communicate and interact with the player in a limited way.

Our experience so far is generally positive: NWN allows us to do what is necessary. The problems of the inappropriate look can be resolved with more modelling, and the faithful representation of the larger terrain scale would make gameplay difficult anyway. The layout of MacTown being modelled after McMurdo Station clearly seems to lead to incidental learning: while playing gamers, rapidly became familiar with the layout of the base, and the roles of the different buildings.

3.2 The Quests

Quests are the core of the game. In the original NWN, they serve to present the player with puzzles and tactical decisions, and both facilitate guidance and document progress. We hoped to use these structures to aid our educational work. We have developed two main types of quests, active quests, which are essentially task-based, and rhetorical quests based on conversation. Active quests are constructed around the process of scientific exploration, from primary data collection, to making inferences and conclusions based on analysis of the data. Rhetorical quests are more experimental, and we discuss them further later in the paper.

One action quest explores the process of ice core drilling, for determining the climatic state of Antarctica at different periods. The player needs first to collect ice core samples, then to bring the sample back to the station for analysis, and lastly to make inferences on the climate of Antarctica. Quests are structured into sub-quests and plots. Plots are constructed around scientist actions or actions to be taken by players.

Resulting climate changes are illustrated by dramatic and humorous visualisations depicting melting ice and rising sea-levels. The climate changes are illustrated through visits to a small island, which goes from a frozen landscape, to a tropical landscape, to finally being submerged under meters of water. Multi-sensory modalities, such as sound and images, can be combined to evoke powerful emotional images that are remembered.

Another example of an active quest is the "the penguins krill quest" (see figure 1b). In this quest, we want students to understand the impact of global warming on a specialized species, but we want to accomplish this within the framework of enjoyable gameplay. For this quest, we first require that players transform into penguin, as only a penguin can talk to another penguin: we can allow this via scripting. Seeing problems through the eyes of the penguin adds emphasis and makes the quest more personal and humorous.

Conversations between characters, and between characters and players have an important function in our game. Mediation through language is critical to learning [12]. We are designing specific rhetorical quests and plots based on conversation, using different narrative modes and techniques such as humour, to communicate information, challenge students and foster social emotions. Rhetorical plots have an important phatic role in our game, they play a similar role to film transition, they initiate quest and *move* players from one quest to another. To keep players interested, plots need to be designed creatively using different narrative modes. We are also exploring conversation as an alternative to fighting, which featured strongly in the original game.

Figure 1. a, b: screen shots of Antarctic NWN, Mc Town and Penguin quest.

Specific argumentative quests can also be developed following the old Greco tradition, to debate issues, challenge an opponent or raise questions. Reflexivity and dissonance can create powerful learning opportunities for players as they struggle with multiple viewpoints and diverse value perspectives presented by the characters [13]. This can be done seamlessly within gameplay, also enriching the sense of immersion. The support for inspecting quest progress also facilitates reflection and planning.

Our experience so far with the quest structure has been very positive: quests are well supported by the game and critical to gameplay, enable us to further our educational objectives, and allow us to work with affective techniques to integrate the gameplay with the educational strategy.

4. DESIGNING A COMPELLING GAME

Our initial emphasis has been to explore in order to determine whether NWN can support education, so we are in the early stages of the game development. In this section, we want to present a number of design issues like imaginary simulation, emotion and humour, which are particularly relevant to the design of an enthralling game such as Antarctic NWN.

As we have stated, emotion is particularly important in our approach, as emotional energy both drives and holds together the development of the game experience [14]. Engagement arises from the participation of students with the environment as they become involved in a drama. Thus, narrative and emotional elements should be revised to insure the player emotional involvement.

We developed our game from an existing game featuring adventure quests, based on fast actions and vanquishing villains. By contrast our game is more contemplative, still based on role-play, but solving scientific enigma, and resolving conflicts. There are several implications, and we must find different innovative design solutions to maintain engagement.

The issues at stake here are the relation between reality, simulation and game word; and between engagement and learning. If we are to succeed, we have to realise the fun and compelling aspect of games but also to support aesthetic understanding.

4.1 Imaginary Simulation

The fundamental design basis for our game is imaginary stimulation. Antarctic NWN is neither the fictive world of Neverwinter Nights, nor a simulation of McMurdo Station and life in Antarctica. There are important distinctions between a simulation and a game.

Games can be completely imaginary and fictive worlds, but games might also be an artistic representation of a simplified reality such as SIM city [15], which directs and immerses the player in the game. Simulations are based on an accurate model of reality that designers have streamlined to accommodate users' cognitive abilities or to overcome technical limitations. Our design challenge is between fantasy and simulation or, combining vividness and imaginativeness. Vividness allows players to feel the reality of a game while imaginativeness relates to the fantastic aspects [16].

Another game mod based on NWN is *Revolution,* where the setting is the American Revolution [17]. In this game, designers used a historically accurate model of a town and characters' costumes. Although we are initially using the medieval look and feel of NWN for practical reasons, it is by no mean obvious that we should model a more realistic description of Antarctica. Simulating a more "realistic" Antarctica might not necessarily be the best design solution.

In a scientific quest such as the ice core quest, we might want to use a more accurate version of scientific instruments to facilitate learning. The barbaric look of some of the characters could also prove distracting. However, keeping a fantasy look and feel to the game led us to design the king of the penguin quest, which would look odd in a more modern setting. The more fantastic and imaginative aspects of the game also allow for a more creative gameplay. One design challenge that we are facing is how to orient the game between fantasy and simulation.

One important question is thus how much we want to convey of the Antarctic reality and how can we express it within the game. We can for instance further embed specific elements of Antarctic reality in the game that are relevant to some of our target population (i.e. SOI). Antarctic weather is a hazard, so to emphasise the importance of Antarctic clothing and gear, characters should always dress up before going out. We expect that such basic facts about Antarctica can be learned incidentally from playing the game. Moreover, adding danger elements such as extreme cold, generators breaking down, or ice shattering will emphasise hazards, but they can also be used within the strategy of the gameplay as obstacles to overcome or random surprise elements that have to be addressed at once to keep players engaged.

The creation of an imaginary simulation demands that we carefully chose Antarctic issues and elements that we want players to explore and contemplate, and find how to communicate them through innovative design solutions that support the gameplay and maintain engagement.

4.2 Designing with Humour

In this game, we are exploring the uses of humour to make the game engaging. Besides, there are a number of ways that humour enhances learning. Humour can help to manage learners' well-being, emotions and mood, for example, and help prevent frustration, or reduce anxiety. Humour has motivating effects on attention, retention and comprehension of information [18]. And of course using humour also provides a more pleasurable experience.

In the game, the choice of characters' names has often a comical flavour. We see humour as relevant to dialogues to encourage players to read conversations. It has been claimed that humour makes the interaction between characters seem more natural and enjoyable, and humour can play a role remarkably similar to small talk [19]. Humour greases the wheel of dialogue, serves a transitional function, and provides a ritualised way for people to move in conversation.

Puns and other forms of verbal joke can enliven conversation. Humour is used in our penguin quests to relate an anecdote about penguins: the king penguin is the ultimate winner of the game "king of the hill". Irony can also play a role, as this rhetorical device has the ability to evoke a closer relationship based on a shared experience. An important function of humour is to create a link with the audience.

Characters can also be personalised using humour. The study of physical comedies shows that exaggerations of body postures create a comical effect. In one of our argumentative rhetorical quests, the character debates in a very forceful way, and characters can be given very distinctive and contrasting body types (such as in the old comedies) to increase the slapstick effect. Building on the strengths of humour, representations can be exaggerated or transformed to have a comical effect. Building on visual comedies might provide alternatives to fighting villains, and enhance the fun and emotional involvement of players.

4.3 Making Characters Cry

As we have discussed emotional involvement is an important part of designing for aesthetic understanding. Emotion can augment engagement and enhance the pleasure of learning. It has been argued that the chief impediment and motivator to learning is not cognition but emotion [13]. Emotion is especially relevant in a role-play game as it draws players in the story, and makes for richer characters, as well as providing information.

Emotions in games are evoked, for example, by failure and achievement, by obstacles overcome and by success in quests. Emotionally rich environments can be created by carefully choosing lighting, colour, music, sound, objects, symbols and character's conversations and actions. We can darken interiors to represent a heavy atmosphere, and costumes can also indicate variations in moods etc. The sounds of Antarctica, diesel generators at the station, wind, penguin or seal colony noises, these will be added to the game to enhance vividness. We want to accentuate the aesthetic and romantic feel of the game to reflect the fascination for adventurers and explorers like Scott and Shackleton. Characters, actions, dialogues and props all should function to provoke a compelling experience.

In NWN, conversations between characters take place in a chat window or in the game environment. Characters' emotions can be emphasised through typographic variations, and the inner feelings of a character can be represented by using specific symbols, as is the tradition in comics.

Human beings are experts at interpreting facial expressions and body language, and they are profoundly susceptible to emotions, moods and actions of others [20]. Visual persuasion has shown how the use of happy people makes us more cheerful, or how showing excited groups arouse viewers' curiosity [21]. Empathy with film characters keeps a viewer engaged in the film world; similarly empathy with game characters can engage players in the game world. Players would project their own values and desires into the virtual characters.

It could be thus quite important for a role-play game, to develop a strong pallet of basic emotions. Crying can be conveyed by sound effects and by showing the character with his head hanging down and shoulders crouched. We plan to strengthen the characters' facial expressions through simple conventions, e.g. a character turning very white (fear) or red (anger). Facial expressions are not usually shown in great detail in NWN, so it may prove difficult to evoke with precision. However, it is believed that characters' emotional value should be studied in this context and examined in relation with the believability and engagement of the game.

5. CONTEXT OF ANTARCTIC NWN

In our quest to understand how games can both delight and instruct we need to consider in depth the environment in which Antarctic NWN is played. We envisaged that the game would be played at home as school homework however we also want to explore the use of game in a different context, as family entertainment. From the basic game, we discuss some of the design issues related to the context of use.

5.1 Education and Antarctic NWN

Bridging the educative and ludic world might not be simple. From the educational point of view, Kurt Square [22] suggests creating instructional resources around any game, to push students to think about their game-playing more deeply. However, we should be careful when using computer games with an educational purpose not to offset the gameplay and the learning dynamics against each other.

When using Antarctic NWN as homework, we need to support students in their learning process. An interesting component of NWN is the role of henchmen. Henchmen are characters that a player can recruit throughout the course of the game to help them in their quests. A henchman is thus specifically designed in our module as a guide, to provide additional explanations when required and to help students. Learning to play a game is not always straightforward especially for non-gamers. Teachers can play the role of "Dungeon Master" as in the Dungeons and Dragons games. In this way, they can motivate and support students playing at home, for example via a chat window. Role playing games also typically a number of associated manuals (frequently third-party) that are used to assist in role playing and are well suited to educational purposes. In these manuals, we can include additional materials such as references to related websites, or point to classroom related activities. For example, in one of the quests scientists would start to argue about the role of CO2 emissions in global warming. Students could then be asked to research and review the concepts surrounding global warming to prepare for a political debate on this subject in the classroom. Participating actively in debates should enhance the learning experience.

Another more challenging possibility is to give students the opportunity to develop their own stories, and to design and implement their own quests. It could be an important extension activity, and might provide motivated teenagers significant opportunities for knowledge-creation and co-authoring of their own educational game environment.

We have presented some of the issues pertaining to the use of Antarctic NWN as school homework centred on supporting education. Different game elements such as the henchman guide or scientific quests will be tested in relation to their educational content but also in relation to the compelling nature of the gameplay. Learning involves both individual and social processes; learning is stimulated by peer interaction [23]. For more advance versions of Antarctic NWN, we envisage transforming the single player game into a multi-player game.

5.2 Antarctica NWN as family entertainment

The family is not a natural unit that simply exists, but one that is defined by action. Most games are played with friends or family, and video games are no exception. Mitchell gave Atari 2600 consoles to twenty families and found that most families used the game systems as a shared play activity; video games were found to be a positive force on family interactions [24]. Moreover, the interest for computer gaming is changing: around 40 % regular computer game players are adults over thirty and two-thirds of the 'gaming demographic' are female [25]. Thus it is quite possible that an entire family play computer game.

The objective of this version is to explore the potential developing family based educational games. It enables us to look at a different relation between learning and entertainment, and learn more on how to design for a family game in this context.

We can establish our game within a family network, as a multi-player game for family members. Each family member would play a different roles that matches the characters already developed in the game such as scientists, students, the penguin king, and so on. However, in such a family version we would need to more strongly support the social dynamics. Each person would be part of a scenario in which tasks cannot be completed that without the participation of another family member. To enhance the gameplay, we can populate the game with clues about the characters' inner lives, family situations and history. For example, we can introduce the journal of the great-great grandfather explorer, describing Antarctica in his time, and reminiscencing about family history. The character cast needs to be modified to relate to family members and personalities. Studies have suggested that older siblings can play an important role in the introduction of home technology, especially for girls [26]. We can also introduce older helpers in the cast whose mission is to help younger players to solve quests. A grand-fatherly henchman could be introduced to help younger players still. A family of penguins could be developed with a much younger cute but annoying little penguin. In other words, we can use a variety of devices familiar from stories, plays, and films, to assist identification and to allow engagement from several perspectives in a family. For this game version, we also probably need to customise quests and to make them easier for younger siblings and more challenging for older family members.

To tailor the basic Antarctic NWN game as a family entertainment we need to consider a number of issues like game issues such as character cast and quests, as above, but also informal learning and engagement. In our design, we consider that learning will happen incidentally when playing the game, in the way word games support literacy and card games numeracy.

Making the gameplay appealing to all family members might prove more difficult. We plan to use Antarctic NWN as a means to help study the potential for family computer learning games in more depth.

6. CONCLUSIONS

In this paper we described our initial work on re-purposing a computer game with an entertainment focus to also support educational objectives. We worked with the game Neverwinter Nights to address a geography education topic on global warming by developing our game Antarctic NWN. We focused in this paper on fundamental a design issue: how to integrate learning objectives while maintaining the compelling nature of a game. We described how we looked to the structure of the game to find support for our educational objectives. We found sufficient support in the way the game is oriented around a setting and scriptable non-player characters; this allowed us to build and populate a world suitable for learning about global warming.

The game does not easily support realistic terrain modelling or scale, but it does allow us to model layout and character roles to support incidental learning. The game also revolves around quests, and we found the narrative structure of quests in particular quite valuable. We described how we use various action quests to give guidance and support learning about processes. We are also using rhetorical quests to guide collaboration and competition between the player and non-player characters. This also allows us to explore alternatives to fighting which features strongly in the original NWN game.

One of the main objectives of our project is to create an engaging educational game. We focus on imaginary and simulated aspects of the game and how they can affect the aesthetic and educational aspects of the game. The role-play in the game means that character and role identification can be leveraged, thus we more specifically discuss the design of emotional characters. Supporting emotional involvement is an important aspect of the game, thus we emphasise the role of humour, to make the game fun, the character more believable, and encourage learning. We found that the game does allow us to represent emotion by character interaction and posture, but it was more difficult to show details using facial expressions.

In the last part of our paper we looked at the game context of use and how it influences the design of our game. A question that has to be investigated concerns the contextual effect of the classroom, and of the game being homework. This setting could negatively affect engagement, i.e. the game being less fun because it was homework, but it could affect positively learning, by providing a fun and stimulating experience. To explore differences in learning and engagement, and the role of context, we discussed the design of Antarctic NWN as a family digital game.

Our experience suggests that it is indeed possible to support education within the game framework we chose. We are continuing to develop the game iteratively and incrementally, attempting to match the game support structure with our pedagogical objectives, and working with students to evaluate the experience. In particular, we will pay closer attention to differences in gender and between gamers and non-gamers. There are acknowledged differences in what kind of computer games girls and boys like, or what they like in computer games. This could also influence the design of future iterations with the creation of different characters and quests adapted to each gender.

We hope to develop a sufficient understanding to inform the development of compelling educational applications with mod games, and establish a model of the game design process, with guidelines to help future designers.

ACKNOWLEDGMENT

This work was made possible by a fellowship from Cognos Inc. The Cybercartography and the New Economy project for which Dr. D. R. Fraser Taylor is the Principal Investigator is funded by the Social Sciences and Humanities Research Council (SSHRC) of Canada under the Initiative on the New Economy (INE) Collaborative Research Initiative Grant.

REFERENCES

[1] C. Phillips, S. Rolls, A. Rouse, M. Griffiths: Home video game playing in schoolchildren: a study of incidence and patterns of play. Journal of adolescence, 18, 1995, pp. 687-691.

[2] D. Gentile, P. Lynch, J. Linder, D. Walsh: The effects of violent video game habits on adolescent hostility, aggressive behaviors, and school performance. Journal of adolescence, 27(1), 2004, pp. 5-22

[3] Young, J. R. Upitis The microworld of Phoenix quest: social and cognitive considerations. Education and Information Technologies 4(4), 1999 pp. 391-408.

[4] F. Taylor: The Concept of cybercartography, in Peterson, M. (ed) Maps and the Internet, Amsterdam: Elsevier, 26, 2003, pp. 405-420.

[5] British Antarctic Survey, 2004, http://www.antarctica.ac.uk/About_Antarctica /FAQs/faq_02.html

[6] D.McGonigal, L. Woordworth: The complete encyclopaedia: Antarctica and the Arctic. Firefly Books, Willowdale: Ontario, 2001.

[7] S. Mumtaz: Children's enjoyment and perception of computer use in the home and the school. Computer & education 36, 2001, pp. 347-362.
[8] S. Egenfeldt-Nielsen: Practical barriers in using educational computer games. On the horizon, 1, 2004, pp. 18-21.
[9] J. Meredith, R. Fortner, G. Mullins: Model of affective learning for Non-formal Science education facilities. Journal of research in science teaching, 34, 8, 2004, pp. 805-818.
[10] M. Girod, C. Rau, A. Schepige: Appreciating the beauty of science ideas: teaching for aesthetic understanding, Science education, 87(4), 2003, pp. 574 - 587.
[11] Bioware, 2003, Neverwinter Nights, http://nwn.bioware.com/
[12] L. Vygotsky: Thinking and speech. In the collected works of L.S. Vygotsky. Plenum press: New York 1, 1978, pp. 121-166.
[13] A. Pedretti: Perspectives on learning through research on critical issues-based science center exhibition, (2004).
[14] D. Wong, and K. Pugh: Learning Science: a Deweyan perspective, Journal of research in science teaching 38, 3, 2001, pp. 317-336.
[15] SIM City, 2004, http://simcity.ea.com/
[16] D. Choi., H. Kim, and J. Kim: Toward the construction of fun computer games: differences in the views of developers and players. Personal technologies, 3, 1999, pp. 92-104
[17] Revolution, the game prototype, 2004 http://www.educationarcade.org
[18] N. Kher, S. Molstad, and R. Donahue: Using humour in the College classroom to enhance teaching effectiveness in 'dread' courses. College student journal, 33(3), 1999, pp. 400-407.
[19] A. Nijholt: Embodied agents: a new impetus to humour research. In Proceedings of the 12th Twente workshop on language technology in computational humour, April 15-16, Italy, 2002, pp. 101-112.
[20] G. Dyer Advertising as communication. Routledge: London, 1992.
[21] F. Caudle: Communication and arousal of emotion: some implications of facial expression research for magazine advertisements. In emotion in advertising eds S. Agres, J. Edell and T. Dubtsky. Quorum books: London, 1990, pp. 127-159.
[22] Squire K. Reframing the cultural space of computer and video game, 2003. http://cms.mit.edu/games/education/research-vision.html
[23] J.P. Gee: What video games have to teach us about learning and literacy. MacMillan: New York, 2003.
[24] E. Mitchell: The dynamics of family interaction around home video games. Special Issue: Personal computers and the family. Marriage and Family Review 8(1-2), 1985, pp. 121-135.
[25] J. Dyck,, D. Pinelle, B. Brown and C. Gutwin: Learning from Games: HCI Design Innovations in Entertainment Software. Proceedings of Graphics Interface 2003, 2003.
[26] J. Fromme: Game studies: computer games as part of children's culture. Game Studies, 3 (1), http://www.gamestudies.org/0101/juul-review/

SUPER ASSIST

Personal Assistants for Diabetes Healthcare Treatment at Home

Geert de Haan, Olivier Blanson Henkemans, Mark A. Neerincx and Charles A.P.G. van der Mast
Man-Machine Interaction Group, Delft University of Technology, Delft, Netherlands. O.B. Henkemans and M.A.Neerincx also at TNO Human Factors Research Institute, Soesterberg, Netherlands.

Abstract: This paper describes the research plans in the SuperAssist project, introducing personal assistants in the care of diabetes patients, assisting the patients themselves, the medical specialists looking after the patients' healthcare, and the technical specialists responsible for maintaining the health of the devices involved. The paper discusses the issues of trust and cooperation as the critical success factors within this multi-user multi-agent (MUMA) project and within the future of agent-based healthcare attempting to increase the self-help abilities of individual patients.

Key words: Human-computer interaction, personalization, supervision, ambient intelligence, health care, personal assistants, cooperative problem-solving.

1. THE SUPERASSIST PROJECT

TNO Human Factors (TNO), Delft University of Technology - Man-Machine Interaction group (TUDelft) and Leiden University Medical Centre - Medical Informatics (LUMC) are developing guidelines, models and methods for joint user-"electronic assistant" supervision of critical equipment and information.

The aim is to establish effective and efficient distributed supervision of networked information compilations and technical equipment, which is trustworthy for the user and takes place in a socially approved manner. Specific innovative project results are: communication and interaction model for these assistants; methods for joint human-computer supervision; improved test methods, tools and criteria for systematic assessment of user experience; "best practice" implementation-method and guidelines; and a "proof of concept" in the transmural health care domain (e.g. diabetes).

For the medical application domain, the SuperAssist framework aims to reduce the costs by improving the local, self-care capacity of people by efficient employment of remote, distributed expertise. Figure 1 provides a possible setting for the SuperAssist framework in which both, a patient, the medical specialists who treat the patient, and a technical specialist who looks after the proper functioning of the medical equipment involved are all provided with personal assistants to facilitate their own tasks as well as to support cooperative tasks which might occur, for instance, with the introduction of new drugs and therapies or with the inevitable malfunctioning or erroneous use of the medical equipment.

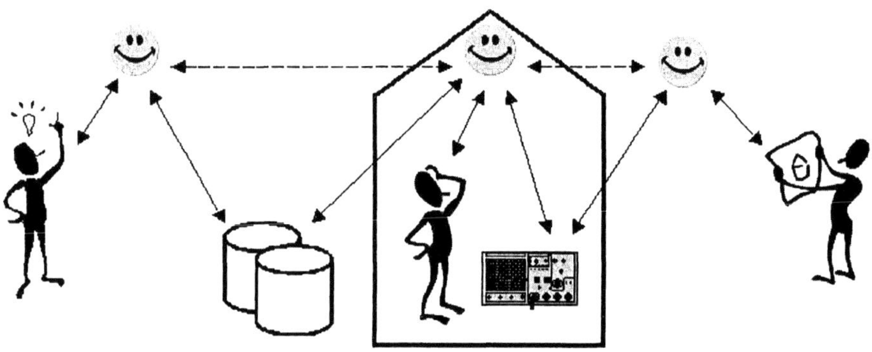

Figure 1. Example situation in which all parties involved are supported by personal agents

The project's business partners - Science & Technology, Philips Research, Pemstar and Sigmax PDA Solutions - bring in their technology and contribute to the development and validation of SuperAssist elements. The joint activities are included in cognitive engineering cycles, in which the foundation, specification and demos of the SuperAssist concept are being refined and extended. The SuperAssist research takes place in the medical domain, but it aims at a generic solution for the distributed supervision of complex environments.

2. A USER SCENARIO

Ms. Brown is a vital 72 year old who, after losing her husband several years ago, refound her balance and takes a keen interest in bicycling and reading. Two years ago she was diagnosed with type II diabetes for which a diet and medication were prescribed. Although she is sincere in following up the medical directions, sometimes her memory fails her and she simply forgets things like when she has last taken her pills or that she should only buy low glucose products. Like many of her friends she is enjoying life but every now and then coping with the restrictions of old-age takes the form of a struggle. Although she accepts having an incurable disease, she is definitely not going to live her life like a sick person.

Since ms. Brown is well overweight, she has been asked to participate in a therapeutic program whose purpose is to attempt to let participants lose some of that weight in a responsible yet playful manner. As a result, every day before breakfast and before bedtime Ms. Brown consults her personal diabetes assistant, provided to her as part of the program, to check her blood glucose levels and to help her remind whether she has taken her pills or not. She can also fill in a questionnaire regarding food intake, exercise, and stress in order to help her manage her food intake in relation to her activities. Since ms. Brown does not to be her illness she has kindly refused to carry the assistant all day as a portable dietary advisor. Because the diabetes assistant is not used as a portable and to address her deteriorating eyesight, the assistant is equipped with an extra-large extra-bright screen, a few large and clearly labelled buttons. In addition, the form and content of the dialogue have been adapted for ms. Brown according to her preferences and to the way she uses the diabetes assistant.

On three consecutive days, ms. Brown's blood glucose level has been slightly higher than normal and today it is rather high. An abnormal reading is not necessarily a cause for alarm. One possibility is that there is nothing wrong with ms. Brown but rather that the technique let her down. The reading of the blood glucose level is derived from the light passing through the finger that ms. Brown offered for testing, which is much more comfortable but also less reliable then blood sampling. Ms. Brown may call the district-nurse to do blood tests but she prefers not to, because he is busy enough with the "really sick people". Finally, the measurement might be genuine. It may be that ms. Brown has forgotten to take her night time pills. However, the display of the assistant clearly shows that she has confirmed taking her pills until this morning and besides, the compartments of her pill-box are empty. It may be -again- that she has accidentally taken a regular instead of a low sugar desert.

A little alarmed, ms. Brown at first doesn't know what to do but then she sees the help-button on the diabetes assistant and remembers from the instruction session that "if you would like to ask a question, first try the help-button". This she does and a friendly voice assures her that there is only a little worry but not an acute problem. Furthermore, the assistant suggests her to redo the measurement, this time using the little finger of her other hand. Using her other little finger, ms. Brown now learns that her blood glucose level is only slightly higher than normal and her assistant asks her to take her pills, including an extra TZD "you know, the big blue one" just to be on the safe side.

3. HIDDEN SCENARIO ASPECTS

By this time, also behind the scenes some activity has taken place. Because of the measurement error, a maintenance module inside the assistant increased the problem count of the measurement module, reset it and performed a self-evaluation test in order to exclude a range of technical problems. Also a temporary wireless network is set-up for a number of routine and problem reports. When the assistants' problem count exceeds a certain level of serious measurement or transmission problems, a technical expert is informed for further maintenance or for a replacement of the device.

In addition, with serious problems the district nurse will be informed of the existence of a possibly faulty device, so that he can check out to take over or provide a replacement diabetes assistant.

Because in this case, the measurements remained within the safety boundaries, there was no need to alarm the specialist diabetes team, although, as a matter of routine, both blood glucose readings and the measurement problem are fed into ms. Browns medical dossier.

Finally, because there has been a change in ms. Brown's otherwise rock stable blood glucose level that deviated from normal over a number of days, a reminder is send to the district nurse. The personal patient-visiting assistant agent will remind the nurse the next time he goes on a periodical visit to ms. Brown. Even though the diabetes assistant is "allowed" within certain boundaries to change a patient's medication, only medical specialists (or in their place the district nurses) are allowed to make any enduring or substantial changes.

DIABETES TYPE II

Diabetes type II or "old-age diabetes", occurs when the body doesn't use the available insulin effectively. Because of the low insulin level, the blood glucose level may build up, causing a sick feeling. Type II diabetes is generally caused by a life-long combination of a bad diet, lack of exercise and overweight. The treatment of diabetes II is focused on keeping a healthy blood-glucose level, which should neither be too high not too low. If the glucose level drops too low, there is the possibility of passing out, coma and eventually death. A minority of diabetes II patients has to inject themselves with insulin (about 15%) but for most patients the treatment consists of a diet, in combination with medicines to boost the production and the effectiveness of insulin.

Diabetes II is not deadly by itself but its symptoms, in the form of a deterioration of small blood vessels may lead to blindness, numbness, inflammations, etc. which may eventually cause death. In principle, diabetes II is not difficult to treat. However, because of the life-long bad eating habits and the much increased feeling of wellbeing by medication, it is often difficult to get sufficient cooperation from the patients well before the point when the bodily condition really deteriorates.

Since most diabetes II patients are elderly - though starting from their forties - feeling secure about the illness and the care they receive tends to be more important then in other diseases. As such, patients should not just be told how to live, what to eat, and which medicines to take how often, but they should also receive specific reassurance about the fact that one may grow old with diabetes.

4. CRITICAL SUCCESS FACTORS IN SUPERASSIST AND THE FUTURE OF HEALTHCARE

In the SuperAssist project we intent to follow a scenario-based approach [1] in that the project consists of a number of phases, each of which is approached iteratively, and guided by hypothetical and real scenarios of would-be situations [2]. In dealing with the novel situation of having multiple types of users communicate and cooperate with multiple types of agents, the domain is relatively new and, according to Erickson [3], a scenario-based approach is the preferred method. In addition to such standard HCI practices, it is a "sine qua non" that users should participate in the development of a personal service or consumer product that SuperAssist aims at [4].

Even though the project targets at a novel type of application in a new area of multi-user multi-agent (MUMA) interaction, it does not start from scratch. Apart from being concerned with software agents which assist human beings in the area of ubiquitous computing [5, 6], SuperAssist attempts to build further upon the PALS project, a Dutch project investigating the requirements to meet changing user needs and usage contexts with respect to web-services (see:. [7, 8, 9]). From this and other projects, two factors turned out to be of critical importance for the success of human-agent interaction: trust and cooperation. Trust in the sense that people need a basic level of trust in a system or personal assistant to "do business" with them, and cooperation in the sense that, like in human-human cooperation, participants rely on images, models or ideas about the communication and cooperation abilities of their discussion partners in order to set appropriate expectations.

4.1 Trust

The trust of a patient, a medical specialist or a nurse in their personal assistant is a prerequisite to delegate part of their responsibilities to the electronic device. To enable trust and delegation between people and electronic assistants, it is a first necessity that the form and content of the interaction dialogue are acceptable to the human user, including a possibly frightened patient or an excessively busy medical worker.

Here, it may be possible to adapt the dialogue to the current context of use including aspects like fear and stress [9].

Apart from a humanly usable dialogue for solving problems, it is necessary that there is a shared view on the problem domain. This may be achieved by creating a representation of the problem space in which rules exist or may be derived to guide the behaviour of the assistants. Note that, even in a well-delimited problem domain as diabetes, this is a major undertaking due to influences from outside the system, such as the nurses' other responsibilities or the complex (side) effects between things like medication, stress and diet.

4.2 Cooperation

Cooperation for problem solving between people and their assistants and the cooperation between assistants among themselves requires some kind of model which at least describes 'what to expect from whom' in terms of questions, actions, etc. In (software) multi-agent architectures like e.g. FIPA (see: [10]) the question 'what to expect from whom' is specified in agent-communication models, either by way of static models, specified beforehand, or in on-request models, used in dynamically changing environments. FIPA is a proposal mainly aimed at enabling communication between software agents but for heterogeneous networks with human agents, it is similarly required that some model exists to specify what other participants may contribute and how to ask them; either directly or by means of some user-agent. In the scenario this is exemplified by ms. Brown's diabetes assistant asking her (on behalf of some supervision agent) to redo the measurement of her blood glucose. Also here, personalisation and adaptation come into play, for instance in deciding whether is appropriate to ask the patient to do something or to alarm the nurse. In addition to this it is probably even more important to utilise inter-agent communication to prevent the human participants from becoming overloaded with questions from their software counterparts.

5. PROJECT PARTNERS AND RESEARCH PERSPECTIVES

In terms of finance and personal, SuperAssist is a relatively small project. Nevertheless, as a project, the ambitions are rather high, if only because of the way in which all of the participants complement each others knowledge and experience: in the area of self-healthcare at home, each partner contributes a piece that is both necessary and unique to find solutions to the problems at stake. In this context, Science and Technology (S&T) provides their knowledge and experience in health management of electronic devices in areas like aerospace and space travel. In the SuperAssist project, Pemstar and Sigmax PDA solutions apply their engineering and PDA adaptation experience to non-professional and healthcare users for personal assistance. Philips Research provides their HomeLab usability evaluation facilities to evaluate the results of the project in the context of healthcare at home as well as to complement their own Ambient Intelligence research programme in which self-healthcare is assumed to be a major selling argument [11].

To the research partners, TNO Human Factors (TNO-HF), Delft University of Technology (TUDelft) and Leiden University Medical Centre

(LUMC), SuperAssist provides an opportunity to extend their knowledge guided by a practical problem. As such, the TUDelft is most interested in investigating architectural issues required for MUMA systems. Earlier in this paper, multi-agent architectures like FIPA [10] were briefly mentioned. A main research question concerning SuperAssist is which agent-architecture is most proper for MUMA systems, and similar to the fat/lean client discussion in client-server systems, how 'fat' (or alternatively, how 'lean') the personal assistants' agents should be to enable, both, a meaning interaction with the human user and a functional interaction with other agents, particularly since SuperAssist's personal assistants should reside on machines like PDA's with limited computing and energy resources.

The perspective of TNO-HF is focussed on MUMA cooperation and mutual problem solving in non-professional areas like self-healthcare. In addition to TNO's existing experience in user-agent cooperation in military and transportation applications, SuperAssist creates an opportunity to investigate the application of multi-agent systems in the areas of personal and self-healthcare involving non-professional users with limited memory- and other cognitive capabilities besides a general lack of experience with information technology.

Finally, the perspective of the LUMC is focussing on the combination of Medical and Informatics. In this area, most research has taken place on how to support the medical staff in the treatment of patients. Relatively new in this area is the shift towards programmes for patient self-healthcare, using the internet, for instance (cf. [12]), and the increased use of information appliances for both, patient-specialist cooperation as well as patient self-healthcare. In the

Telemedicine-Diabetes project at the University of Oxford [13] for example, diabetes patients do not merely monitor their blood glucose level to treat themselves, but by additionally phoning-in the measurement results to a diabetes treatment centre, these can be stored in a database and used for personal advice. From these and related investigations, a general "lesson learned" is that patients are generally able and willing to take participate in such mutual medical specialist-patient cooperation project, but only if the personal benefits to the patient are clear.

Hence, when advice and treatment are personalised to the patients context and circumstances. As a result, as part of the SuperAssist project, a main research idea of LUMC is to use a PDA-like device to present diabetes patients with an existing paper "diabetes manual" ([14], to be translated in Dutch) in a personalised, adaptable and context-sensitive way to help them cope with their disease and the often associated overweight. In this context, it is interesting to note that since diabetes patients are somewhat notorious in not being able to change to a healthier lifestyle and diet. As such, the treatment of diabetes in the SuperAssist project should be regarded as a very conservative test, which if it succeeds, should also provide a rather robust solution for other self-healthcare treatments.

6. SCENARIO-BASED DESIGN AND PROTOTYPING TO DRIVE RESEARCH

6.1 Scenario-based Design

Within SuperAssist, scenario-based design [1] will be the main technique to develop research ideas in the project. Scenario-based design will be used in combination with prototyping as the projects main technique to test the applicability and the usefulness of the ideas in more-or-less tangible form. In combination, scenario's and prototypes and mock-ups form cognitive engineering cycles in which they are connected by different levels of specification. Since the project is in an early stage, the main focus is currently on the use of scenarios for two different purposes.

First, scenarios are used to collect 'wild ideas' to envision the outcomes of the SuperAssist project in the form of integrated real-world situations. Scenario-envisioning is thus used to provide some -however idiosyncratic and erroneous- idea of what the project is going to deliver eventually. In addition, by doing so, the scenarios also provide a means to create a common as well as shared understanding among the project members. The first scenario that was discussed in this paper and figure 1 present the use of scenario's for envisioning purposes.

Envisioning future situations is not sufficient to drive a research project. In addition, techniques are required to transform envisioning-scenarios into actual research questions, very similar to deriving research hypotheses from a general theory. To this purpose, the SuperAssist project also uses scenario's at a more detailed level, sufficiently detailed, first, to distinguish the crucial research questions and, later on, to develop research questions at a level that they may be 'embodied' in mock-ups and prototypes. The use of scenario's to focus thinking on which research questions are essential to the project is exemplified in the following scenario.

> Ms. Blair is a 79 year old widow who is still adventurous and consequently, often en route. Occasionally, she suffers from a lapse of memory, causing problems in regularly taking her prescribed drugs. Ms. Blair's physician provided her with a constrained medicine box in combination with a portable interaction device which monitors her

medicine usage and provides for GPRS wireless connectivity. Through the wireless personal device ms. Blair is as it were continuously connected with her physician, her medical dossier, the pharmacist, and a technician. With her 'medicine reminder', ms. Blair is relieved of confusion because only the right pills are accessible. In addition, she is alerted when she runs out of her medicine and subsequently directed to a pharmacist nearby where her new pills are ready to be picked up.

Unfortunately, one day the lid of the medicine box got stuck. A situation like this requires collective supervision from both, ms. Blair, her physician and the technician. As a result, when the personal assistant determines that the lid of the medicine box refuses to open, it sends all three actors an error message. Using her personal assistant, ms. Blair contacts the technician about the specific lid malfunction. In turn, the technician (or his personal assistant) presents a videoclip to ms. Blair, with instructions how to cautiously open the lid by hand. Meanwhile, the physician has checked both the error message and ms. Blair's medical dossier. Finally, he contacts ms. Blair personally to determine that she didn't miss her medication. If this would have been the case, he would prescribe ms. Blair with new drugs which would automatically be implemented in the medicine box.

Important for such a smooth course of joint user-assistant supervision, presented in the scenario is successful interaction. Elaborating on this prerequisite, the following research question can be posed:

Which interaction processes occur between the actors during the joined supervision and how are they task and role related?

These interaction processes will heavily lean on the fine-tuning of the assistant's behavior and the actor's individual needs. In addition, identification of the personal characteristics of the different actors, as users of an interaction device, is requested:

What data about the user, usage and the environment data are needed for a personalized assistant?

How should the assistant acquire and apply such data to realize unobtrusive assistance for the different types of users involved in performing a joint supervision task?

6.2 Cognitive Engineering and Scenario-based Design

Research and design in the SuperAssist project will make use of cognitive engineering cycles, an engineering approach to HCI design with a focus on the human abilities and limitations as the most critical of all design aspects. In relation to other HCI design approaches this means that design is user-centred, involves user participation and takes the need for design iterations for granted. This is not all too different from the common HCI approach to design, except perhaps that an engineering stance is taken in that both, a particular problem has to be solved and that the solution should be systematically applicable to a whole class of similar problems. In contrast to, for example, Dourish [15] who argues that it makes little sense to try to specify beforehand what is relevant in the context of a particular ubiquitous computing applications because context is an emergent property of human-computer systems, a cognitive engineering approach would argue that Dourish may very well be right, in principle but that, in practice, it is may very well be possible to find a solution that satisfies all stakeholders.

The cognitive engineering framework in one in which practical theories, communication and interaction models, and prototypes are being developed in a structured iterative process [16]. A practical theory includes accepted features of cognition and affection, distinguishes different task contexts within a specific application domain, provides descriptions of computer behavior, and predicts the interaction effects between these features, contexts and computer behavior. An interaction model formalizes and quantifies these interaction effects and is used to design a prototype of an "intelligent" user interface, which adapts its behavior to the present goals state, and to the cognitive and affective processes of the user. In addition to the individual user-assistant cooperation model, proposed in [16], the SuperAssist project will develop a co-operation model for the overall, multi-actor supervision, which formalizes the information exchange between all of the actors involved, both agents and men.

The research in SuperAssist is further organized around three main experiments or rather three series of experiments to investigate the main research questions in the project: task allocation among agents and men, the organization of joint anomaly management or problem solving, and finally, the "look & feel" of the assistants, respectively. In addition, a number of flanking experiments and scenario exercises is planned, intended to delimit the design space for the main experiments by excluding in an early stage the less- or irrelevant factors and to find solutions for design details for which a full-size experimental setup is overkill.

Apart from empirical validation of models and theories, the three experiments also serve to validate and improve the design of the prototypes which are designed to run these experiments. The prototypes are used as vehicles to develop both the required knowledge about human-agent cooperation as well as drive the actual design of the specific solution to the problem at hand. As such, in the course of the project, the design of the prototypes being tested changes and becomes more refined and detailed. Whereas the initial experiments will take place using mock-ups, at the end of the project the aim is to present and test a demonstrator in the form of a hafiz prototype. It should be noted, however, that the primary aim of SuperAssist is to acquire and validate knowledge to design MUMA applications. As such, on the one hand, the creation of mock-ups, demonstrators and prototypes is a vehicle to facilitate research; not a 'deliverable' as in many EC research projects. On the other hand, the prototypes serve play the dual role, to validate the underlying theoretical knowledge as well as that they themselves are tested as solutions to the specific design problem.

Having discussed the use of scenario's in general, the cognitive engineering framework, the experimental validation and the use of prototypes, there remains the question how scenario's are specifically used within the framework. How scenario's are used in each stage of design and investigating is exemplified by Rizzo and Bacigalupo (2004) who describe a full range of using scenario's from early design stages, like familiarization and discovery of the main issues to later stages involving validating of ideas and design proposals with stakeholders, creating specifications for design and implementation, and validation of design solutions. According to Rizzo, particularly in new design contexts, it is important not to get lost in the false security of formal specifications. Instead they propose to use scenario's throughout design for envisioning, specification and validation.

In the early stages, the design process starts with observing the relevant aspects of the 'old' design situation, laid down in a number of scenario's or storyboards. Such descriptive scenario's are subsequently used to discuss and develop a common view with stakeholders what the essential problems and principled solutions may be. Note that this stage, like any of the following may require a number of iterations to arrive at a suitable set of working scenario's. In the project, a number of interviews and observation sessions are planned with both support engineers, patients and medical personal.

Next, the chosen scenario's are further analyzed to derive a number of so-called micro-scenario's in which for each part of a scenario, a number of essential elements is identified. This step is most similar to classical task-analysis but instead of aiming at fully describing the stakeholders' tasks, the aim is to arrive at a level of description which allows one to develop and feed-back or validate partial design solutions. The analysis step mostly involves desk work, but to evaluate the feasibility of each of the partial design proposals required cooperation with the stakeholders who actively have to decide which solutions will in practice make sense and which do not. It is at this step that the experiment about task-allocation will be planned and executed.

After analysis follows a synthetic step in which a number of design solutions is taken together as proposals to solve the underlying design problem. The results of this step are called macro-scenarios which describe a number newly proposed task-scenario's in combination together with the supporting technology. Describing design proposals in scenario form rather then in some formal way, makes it easier for stakeholders to be involved. Also here, stakeholders involvement is required, not as individual stakeholders but rather as a community of stakeholders. In such planned feedback sessions, including the planned experiment on joint human-agent problem solving, the role of stakeholders is both to evaluate and select (some) overall design solutions as well as to propose improvements to the selected proposals.

Finally, after the subsequent creation and implementation of one or more of the proposed design solutions, one or more concluding evaluation sessions are held, guided high-level scenario's from the early stages, to validate the overall design solutions and, where necessary, to improve them at a more detailed level. This step results in some form of a final design solution which can be evaluated in the last experiment, concerning the "look & feel" of the design.

It may be clear from the preceding description that scenario's seem particularly useful vehicles to enable user- or stakeholder participation during all the design steps. In addition, scenario-based design approaches also seem particularly useful in situations in which design does not merely involve the creation of a computer application but rather the design of solutions that involve both men and machines -or rather- men, agents and machines with each specific roles and responsibilities to fulfil. What is not yet clear at this stage of the project is how exactly to fit in the planned experiments into a scenario-based approach and to establish a 'best practice' for experiment-backed scenario-based design.

7. ENVISIONED RESULTS AND OUTLOOK

The SuperAssist project attempts to set up an integrated healthcare service in the area of diabetes treatment, assisted by electronic devices and software agents. From this research we expect to be able to derive models, architectures, guidelines and general design knowledge to introduce human-agent systems that enable people to help themselves to a greater extend than they are able to do, now. As such, the SuperAssist case is intended as a test case for a range or healthcare and other services and not just as yet another smart solution for a particular problem. Only by introducing, testing and validating general tools, we will be able to help keep future healthcare costs within the boundaries of affordability and manageability - at the very least for individual patients like ms. Brown and ms. Blair.

The SuperAssist project also employs diabetes as a vehicle to learn general lessons about how to introduce trustworthy and cooperative electronic or software assistants into the everyday life of real people. Until recently, human-agent interaction has been studied in settings where a single agent assisted a single user, as in supporting searching, information filtering, and critiquing systems. In the SuperAssist project, research and development are extended to MUMA or multi-user multi-agent communication and cooperative problem solving. This novel research area does not only introduce new types of questions and new types of problems but it is also a starting point for the integration of separate intelligent applications into intelligent overall services.

REFERENCES

[1] Carroll, J.M.: Scenario-based Design: Envisioning Work and Technology in System Development. John Wiley, New York (1995)

[2] Rizzo, A and Bacigalupo, M.: Report on testing of full scale scenarios. Deliverable 3.2 of the CARE/Innovative Action - CREA Project. Available from: www.dblue.it/pdf/crea/CARE as CREA_UNISI_WP3_D3.2_v1.1.pdf (2004)

[3] Erickson, T.: Notes on Design Practice: Stories and Prototypes as Catalysts for Communication. In: Carroll, J.M. (ed.): Scenario-Based Design: Envisioning Work and Technology in System development (1995) 37-58

[4] Kemp, J.A.M. and van Gelderen, T.: Co-discovery Exploration: an informal method for the iterative design of consumer products. In: Jordan, P.W., Thomas, B., Weerdmeester, B.A. and McClelland, I.L. (eds.): Usability Evaluation in Industry, Taylor and Francis (1996) 139-146

[5] Maes, P.: Reflections on ... Agents that Reduce Work and Information Overload. In: Maybury, M.T. and Wahlster, W. (eds.): Readings in Intelligent User Interfaces, Morgan Kaufmann (1998) 525-536

[6] Weiser, M.: The Computer for the 21st Century. Scientific American, 265(3) (1991) 94-104

[7] PALS, Personal Assistant for onLine Services. IOP/MMI project website; see http://www.tm.tno.nl/pals

[8] Herder, E.: Utility-Based Evaluation of Adaptive Systems . Proceedings of the 2nd Workshop on Empirical Evaluation of Adaptive Systems at the 9th Intl. Conference on User Modeling - Johnstown PA, (2003) 25-30

[9] Neerincx, M.A. and Streefkerk, J.W.: Interacting in Desktop and Mobile Context: Emotion, Trust and Task Performance. Proceedings of the first European Symposium on Ambient Intelligence (EUSAI), Eindhoven, The Netherlands. Springer-Verlag (2003)

[10] FIPA, the Foundation for Intelligent Physical Agents. See: www.fipa.org

[11] Aarts, E. and, Marzano, S. (eds.): The New Everyday: Visions of Ambient Intelligence, 010 Publishing, Rotterdam, The Netherlands (2003)

[12] Alpay, L.L., Toussaint P.J., Ezendam N.P.M., Rövekamp, T., Graafmans, W., and Westendorp, R.: Easing Internet Access of Health Information for the Elderly Users. Health Informatics Journal, 10 (3), (2004) 185-194

[13] Tarassenko, L., Gibson, O. J., Hayton, P. M., Cobern, W. R., Farmer, A. J., Hannaby, K., Dudley, C. and Neil, A.:Mobile Phone Technology to Support the Self-management of Diabetes, Diabetes UK Annual Professional Conference, 17-19 March 2004, Birmingham, UK (2004) (see also: http://www.tve.org/ho/doc.cfm?aid=1589)

[14] Stewart, R. (ed.): The Diabetes Manual, Warwick Diabetes Care, University of Warwick, UK (2004)

[15] Dourish, P.: What we talk about when we talk about context. Personal and Ubiquitous Computing, 8(1) (2004) 19-30.

[16] Lindenberg, J., Nagata, S.F. and Neerincx, M.A.: Personal Assistant for onLine Services: Addressing human factors. In: D. Harris, V. Duffy, M. Smith & C. Stephanides (ed.): Human-Centred Computing: Cognitive, Social and Ergonomic Aspects, Erlbaum, London (Uk) (2003) 497-501

INSTANT MESSAGING

Using log files to identify patterns of interaction in a group of teenage users

Brian Davison and Kathy Buckner
Napier University, Edinburgh, U.K.

Abstract: This paper presents preliminary findings on teenagers' use of instant messaging (IM) as an example of an Internet technology. The results are quantitative, being drawn from the information held in users' IM log files. The current purpose is to validate the proposed data collection method, and to explore the analytical possibilities it presents. This is done through the discussion of a number of graphical representations generated from the data.

Key words: Instant messaging, emoticons, social networks.

1. INTRODUCTION

Today's children are often considered the 'net generation' who will be the inheritors of the true 'information society' (Williams, 1999). However, despite their growing use of the Internet - around half of 9-19 year olds spend between half an hour and an hour online every day - this use is narrow and skills are likewise limited (Livingstone and Bober, 2003; Schiano *et al.*, 2002). Young people's experience of Internet technologies, and the skills they bring with them when they go on to the workplace or to higher education need to be understood in order to be further developed and exploited (Smith, 2002; Farmer, 2003).

When teenagers use online communication tools in their own time and within their own peer groups, they do so out of choice. They are not constrained by externally-imposed goals or rules as are the adults in existing workplace studies (Farmer, 2003; Herbsleb *et al.*, 2002; Isaacs *et al.*, 2002; Nardi *et al.*, 2000). Observation of this natural phenomenon may provide insights into system design, communication behaviour and online learning that are not subject to the preconceptions that may hold true in other circumstances.

Teenagers' use of online communications comes at a time when they are developing their own sense of group and personal identity. IM has characteristics (discussed below) that make it more suitable for this kind of exploration than other kinds of online communication, and this may go some way towards explaining its popularity. If we accept that online relationships are just as real and valid as in-person ones (Peris *et al.*, 2002; Suler, 2004), the use of IM by teenagers provides an opportunity to study their development in a non-intrusive way.

The first step towards either of the two general goals outlined above is the development of methodologies for capturing teenagers' IM behaviour. The next few sections provide the context for this study into the use of data from IM log files.

2. INSTANT MESSAGING

Interpersonal communication has been a key ingredient of the Internet since email was first invented in the early 1970s. Within one year of its introduction onto ARPANET, the forerunner of the Internet, email accounted for 75% of all traffic (Zakon, 2004). Email, however, is *asynchronous* - it does not require both participants to be online communicating in real time. This may have its advantages, but it does not foster interactive conversation. This problem was resolved in 1988 with the creation of Internet Relay Chat (IRC) (Oikarinen, n.d.). IRC allowed one user to type a message which would appear *synchronously* on the screen of the recipient. The development of IRC followed the model of existing asynchronous newsgroups, and let to the concept of the *chatroom* - a virtual 'place' where users would go in the hope of finding others to engage in conversation. Chatrooms are typically organised around topics of interest, an arrangement which enables users to find like-minded conversational partners. Like early email, these chatrooms relied solely on plain text, and in order to convey tone or emotion, users had to use uppercase letters, repetition or symbols such as punctuation. Some conventions evolved, such as the use of uppercase to indicate a raised voice, or emoticons – arrangements of text characters representing facial expressions – to indicate tone. More recently, with improvements to graphical user interfaces, Internet bandwidth and network-oriented programming techniques, variations on the chatroom concept have been developed which allow users to present a visual representation of themselves to others through the use of computer-generated characters, or *avatars* (Talamo and Ligorio, 2001; Suler, 1997; Sulake Labs, n.d.). Such systems are tightly associated with a geographic metaphor - users 'go to' particular rooms to see 'who is there'. In the more sophisticated systems, users can even take ownership of a piece of this virtual space, decorate it as a further act of self-expression and invite other users in.

In contrast to these virtual spaces, IM prioritises the ability to locate and communicate with a set of known contacts, referred to as the user's *buddy list*. This subtle shift in emphasis leads to IM being used quite differently from chatrooms (Livingstone and Bober, 2003) and more like other interpersonal media such as mobile text messaging.

Current IM systems, typified by Microsoft's MSN Messenger used by the teenagers in this study, allow a greater degree of visual expression than the original text-only chatrooms, but fall short of the possibilities afforded by multimedia environments. The avatar is reduced to a single image that represents the user on a conversational partner's screen, and the personalisation of a piece of virtual space is similarly reduced to a choice of background pattern or colour of the conversation window. Enhanced versions of early chat conventions such as emoticons persist in IM systems as shown in Table 1.

Table 1. Some emoticon examples from MSN Messenger

	Smile	**:-)** or **:)**		Open-mouthed	**:-D** or **:d**
	Hot	**(H)** or **(h)**		Sad	**:-(** or **:(**
	Left hug	**({)**		Right hug	**(})**
	Vampire bat	**:-[** or **:[**		Red rose	**(F)** or **(f)**
	Red heart	**(L)** or **(l)**		Broken heart	**(U)** or **(u)**
	Red lips	**(K)** or **(k)**		Star	**(*)**

3. COMMUNICATION NEEDS

Workplace IM studies tend to take a goal-oriented approach to communication, emphasising information exchange and coordination of activity over social conversations (Herbsleb *et al.*, 2002; Isaacs *et al.*, 2002; Nardi *et al.*, 2000). Even where the informal nature of IM conversations is highlighted, it is in relation to the accomplishment of work-related activities (Nardi *et al.*, 2000). Adults in the workplace have a set of role-related priorities, and their communication activity reflects these. The comparatively large amount of time that teenagers spend using communication tools such as IM simply cannot be justified in the same way. For teenagers, contact is more important than content (Livingstone and Bober, 2003), and their intense communication activity is a means of developing a sense of personal and group identity (Boneva *et al.*, in press). The exploration of personal identity comes through for example in the study by Talamo and Ligorio (2001) who report on a transnational educational project centred around a multimedia environment where participants are represented by avatars. Students in the project were much more likely to try out the different avatars available than were the teachers. Suler (1998) suggests that the characteristics of IM environments are especially conducive to teenage identity experimentation. The rarefied nature of personal representation and communication in IM means that it is possible to 'deconstruct' one's personality, and present only one facet at a time. Changing avatars can be a way of changing focus from one facet to another, as can changing one's IM name or image. At the same time, teenagers are experimenting with autonomy and strengthening the bonds with their peer groups as a way of learning about their place in the world (Suler, 1998; Boneva *et al.*, in press). As older teenagers become more confident with their own personal identity, their need for online experimentation diminishes (Schiano *et al.*, 2002).

3.1 IM behaviour

A feature of IM that users consistently refer to as attractive is the ability to hold multiple simultaneous conversations (Lenhart *et al.*, 2001; Nardi *et al.*, 2000; Isaacs *et al.*, 2002; Boneva *et al.*, in press). IM allows teenagers to indulge their need for group belonging by connecting with a number of their peers simultaneously, but each of those conversations is also *private* being visible only to the participants (Boneva *et al.*, in press). Self-reported multitasking behaviour differs from one study to another. Schiano *et al.* (2002) report a maximum of five concurrent IM conversations, while Boneva *et al.* (In press) report as many as 16. Multitasking in IM environments allows teenagers the peculiar mix of group belonging and private interaction in a way that chatrooms do not, and several studies have noted a switch in teenage allegiance from chatrooms to IM with the development of improved tools. While the adults in Grinter and Palen's study (2002) characterise public chatrooms as "a waste of time", teenagers' interest seems to wane with experience (Livingstone and Bober, 2003, p.17). Studies in the UK and America have shown that teenagers are generally aware of the dangers of sharing confidences with strangers online. The anonymity of the chatroom provides ample opportunity for deception, and many teenagers avoid them for this reason (Lenhart *et al.*, 2001, pp. 23-4; Livingstone and Bober, 2003, pp. 21-27).

Because IM conversations typically involve two participants, some of the problems encountered in chatroom conversations are less significant. In particular, the two main requirements for coherence noted by Greenfield and Subrahmanyam (2003), namely establishing who is participating in a particular thread and what constitutes a relevant response, are clearly less relevant with only two interlocutors. While their study describes a number of techniques for exploiting the graphical and textual characteristics of the chatroom to maintain coherence, the same functionality available in IM environments is more likely to be adapted for other purposes.

This study aims to explore the potential analytical uses of IM log file data for the capture of interpersonal relationships and communication activity among teenagers, and to find ways of representing that information in order to facilitate more detailed work.

4. METHOD

MSN Messenger provides the facility to log all conversations to a series of files in XML format, making them amenable to manipulation. The data stored includes the content of the conversation, time and date values for all messages as well as the names of the participants. Within the log files, messages are grouped into sessions, each of which represents the activity carried out in a single MSN window from the moment it is opened until it is closed.

Several teenaged MSN users who were known to one of the researchers were invited to take part in the study by activating their logging facility and providing the accumulated data for analysis. Some were understandably wary and declined, but four informants were eventually found. Having obtained signed consent forms from their parents, all IM activity of these individuals was logged for a period of four weeks.

As a trust building measure, and as a protection of privacy, the content of individual messages was not used in this study. It was also felt that the knowledge that their words would be later read and analysed might inhibit the natural behaviour of the informants. At the point the logs were harvested, the message content was hidden by applying a text filter that replaced all alphanumeric characters with 'x'. This preserved the word count of each message, any linguistic devices such as punctuation and emoticons. The XML document structure made it possible to do this for message content while leaving all other details intact including the names used by the participants. Because the content of conversations was removed, and because of the difficulties inherent in trying to identify MSN users from their session names, it was not deemed necessary to seek wider consent from the informants' contacts.

The result of the first data extraction phase was the series of spreadsheets shown in Table 2. The second phase of the analysis was performed within the spreadsheets shown in the table, or in modified copies. A further phase of analysis was the creation of a matrix of connections between IM participants that was used as input into the network visualisation tool, UCINet (Borgatti et al, 2002). This facilitated the production of a single network diagram representing the connections between individuals - the social network underlying the IM activity itself. Social network analysis is most frequently used to analyse data that has been gathered by questionnaires, from observation or from documents. Questions such as "Who are your three best friends?" or "Who do you ask for help with homework?" are often used to elicit information. This type of question can be problematic as accuracy in response depends on an individual 'choice on the day' and willingness to cooperate and reveal personal information (Scott, 2000). Using log file data avoids this limitation since it relates to actual (measurable) interactions between individuals.

Table 2. Main spreadsheet files

Name	msn_names.xls	msn_sessions.xls	msn_messages.xls
Contents	Participant names Identification numbers	Source (informant and file) Session number, start date/time End date/time Number of messages Mean message length Standard deviation Number of participants List of participant identification numbers	Source (informant and file), Session number Date/time Participant identification number Number of words Mean word length Standard deviation

4.1 Limitations

Four main problems were encountered while processing the log file data. The first of these was the lack of a reliable method for relating the variety of names used by an individual over the course of several sessions. This facility had been present in a previous plain-text version of the log file that had been used when planning this study, but disappeared with the introduction of the XML format. To overcome this problem, an additional step was included in the analysis whereby the informants were asked to verify the attribution of names to particular individuals.

Another issue with MSN Messenger was the duplication of log file data. Under certain circumstances, messages from sessions with more than two participants would be logged to more than one file. This issue is discussed below, but was not explicitly dealt with at this stage.

The third problem related to the processing of some of the exotic characters used by participants as part of their names. The awk text processing tool (familiar from Unix) was not always reliable in its ability to handle Unicode characters, and a small number were therefore lost and replaced by question marks in the filtered files. This did not affect the main analysis, since the majority of names were legible and identifiable.

Finally, a large amount of data was produced, and this created issue for some of the tools being used in the analysis. In particular, both Microsoft Excel, and the spreadsheet embedded in UCINet are limited to 255 columns. This was therefore the maximum number of participants that could be accommodated in the network diagram. Some of the participants with very little activity in terms of the number and size of messages were therefore omitted.

5. RESULTS AND DISCUSSION

5.1 Overall characteristics

In total approximately 30,000 messages were identified, spanning 883 sessions. The earliest logged session was on 4/10/2004 and the latest on 10/11/2004. The exact number of sessions was difficult to determine because of the log duplication mentioned above. The difficulty in attributing names to individuals also made it impossible to give an exact total number of participants, but the approximate figure is 180. This includes the informants themselves, their immediate interlocutors, and a number of other individuals involved in IM sessions into which the informants were invited.

In general, the results obtained confirmed those reported by other authors regarding the number of participants in a typical IM session, and the nature of IM messages. The average number of participants in an IM session was 2.11 with a standard deviation of 0.55. Of the nine sessions with more than three participants, eight were between five and seven, with a single outlier of ten. Clearly, the one-to-one conversation is the norm in this set of data.

The number of words per message was very low – 28% of all messages consisted of a single word, and 52% were three words or less. This again confirms the expectation that IM messages are typically very short. Many of these very short messages could be backchannel contributions intended simply to encourage the other participant and to keep the conversation going. A further factor could be the intense use of linguistic shorthand, contractions and emoticons adapted from email, chat and text messaging (Thurlow, 2003). Some of these forms have the function of entire sentences, and therefore may give a false impression of the significance of a very short message.

Looking at the average number of words per message per participant gave a mean of 4.2 and a standard deviation of 5.11 on the current dataset. The higher mean and standard deviation suggest that there is variation between IM users, each having a personal preference for longer of shorter messages. This was not, however, controlled for the number of message generated by each participant and remains a suggestion at this point.

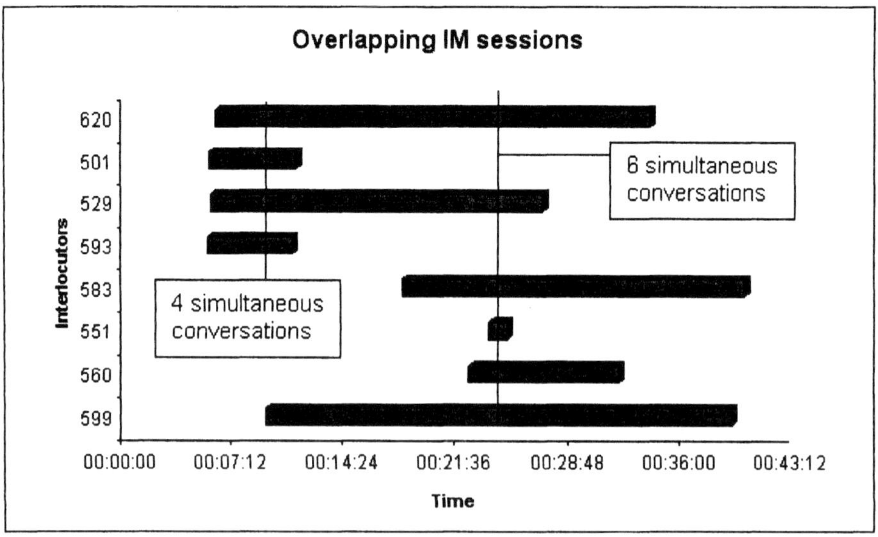

Figure 1. An example of multitasking from a single informant

5.2 Conversational behaviour and multitasking

Comparing the start and end times of sessions from a single informant's log files revealed those sessions that overlapped in time, ie those occasions on which the informant had more than one conversation in progress with different interlocutors. This information was used to produce timeline diagrams such as the example in Figure 1. One model of multitasking behaviour reported by Grinter and Palen (2002) is that of a central group conversation with peripheral one-to-one conversations. Evidence for this would have come from the appearance of group sessions sharing interlocutor numbers with simultaneous one-to-one sessions; however, no evidence for this was seen in this limited set of data suggesting a much less structured arrangement of conversations.

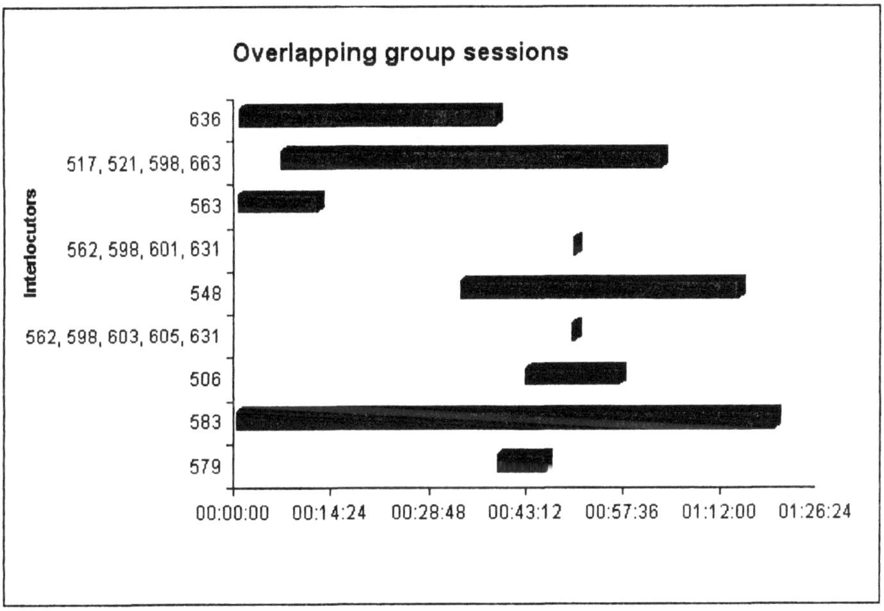

Figure 2. An example of simultaneous group and individual sessions from one informant

In Figure 2, group sessions are indicated by a series of interlocutor numbers on the vertical axis. No group participants are simultaneously involved in one-to-one conversations with the informant. An interesting feature though is the lower pair of group sessions, (562, 598, 601, 631) and (562, 598, 603, 605, 631). These sessions share three participants, and clearly relate to the same message data as indicated by the identical bars on the chart. The reason that they appear as separate sessions is an artefact of the way in which messages may be logged to more than one file by MSN Messenger.

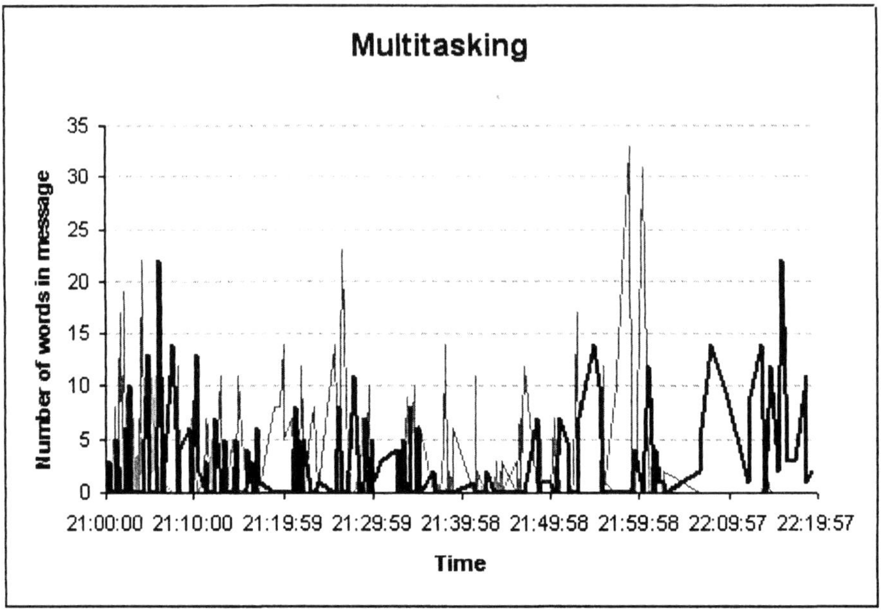

Figure 3. Message activity within simultaneous sessions.

Figure 3 represents the activity of the informant in the sessions shown in Figure 2 in terms of the number, size and timing of messages sent. The heavy line corresponds to the conversation with interlocutor 583 in Figure 2, while activity in all other sessions is shown in grey. Although the highlighted session is the longest, there is no other sense in which it appears to be dominant. Activity is relatively constant all the way through, with some variations where the informant is composing messages to other conversations. This is noticeably the case at the point leading up to the highest peaks on the graph. These long messages – over 30 words each – obviously took longer to compose that the usual rapid-fire contributions, and it is understandable that this should interrupt flow in the other sessions. This suggests, however, that an additional pressure to keep message length short may come from the need to maintain all active conversations. Long messages to one session may cause unwanted delays in others, and a hypothesis for future investigation might therefore be that average message size decreases with the number of simultaneous conversations.

5.3 Identity

Before starting the analysis of the data, it was expected that individuals would use a number of different names. What was unexpected was the scale of this phenomenon. One participant was found to have changed names 119 times during the logging period. This individual was a regular session partner with one of the informants, and was therefore responsible for a large number of messages in the logs. When the number of messages per user was taken into account, it was found that on average a participant changes name after every 22 messages. This, of course, is a very rough figure and does not take into account messages that participants may have exchanged with interlocutors other than the four informants, i.e. messages not captured in the dataset. Nevertheless, it is a good indication of the dynamic nature of identity in IM. There were numerous examples of the participants taking advantage of the visual qualities of the medium, including using emoticons like those in Table 1 to enhance their names. Other typographical creativity involved the use of Unicode characters as in the examples in Table 3.

Table 3. Example MSN names

<table>
<tr><td>1</td><td>§tæ©€¥ []))æ®r€ñ vi©|{í (3îLI¥ €mmæ Jøhñ
(Text parts read as: STACEY, DARREN, VICKI, BILLY, EMMA, JOHN)</td></tr>
<tr><td>2</td><td>ï'm §ùþê®gí®l åñd ï'm hè®ê tõ §åvé thè wõ®ld</td></tr>
<tr><td>3</td><td>(__/)
(='.'=)
(,("'"),)
(")_(") ßït¢hý büÑnÏ catching things and eating their insides</td></tr>
</table>

6. NETWORK ANALYSIS

The representation of the interpersonal links generated using UCINet exhibited the expected explosion of connections surrounding the four informants, shown as light square nodes in Figure 4. Perhaps more interesting though is the information captured about the wider network in which they are located. A number of links between other individuals are identified such as those within circle A, as well as a small group of five in circle B who do not appear related to the main population at all.

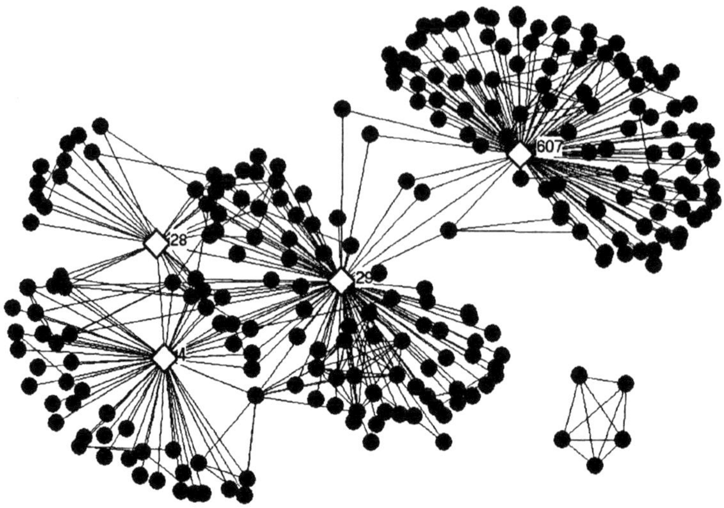

Figure 4. Social network

The nodes in group B are in fact the result of two sessions into which one of the informants was invited, but did not contribute any messages. There is therefore an implied passive link, but only active contributions are represented in Figure 4. A possible line of future enquiry might be to investigate the strength of the links in graphs like the one above, and to test whether passive links are a true indication of peripheral group membership.

Figure 4 also shows that certain individuals such as those in areas C and D hold pivotal positions in the overall structure. The network diagram is therefore an excellent way to identify those people who might profitably be included in a future large-scale study.

7. CONCLUSION

The work presented here has identified a number of opportunities for quantitative analysis presented by IM log file data. It has also identified some limitations inherent in the various technologies used that will need to be overcome in future stages of this study. Future work would need to be based on a larger sample of selected IM users to allow conclusions to be drawn with any confidence, but the value of this kind of information for triangulation with qualitative data is clear.

REFERENCES

Boneva, B., Quinn, A., Kraut, R., Kiesler, S., Shklovski, I. (In press). Teenage Communication in the Instant Messaging Era. In Kraut, R., Brynin, M., Kiesler, S. (Eds). *Domesticating Information Technology*. Oxford University Press.

Borgatti, S.P., Everett, M.G. and Freeman, L.C. 2002. Ucinet for Windows: Software for Social Network Analysis. Harvard, MA: Analytic Technologies.

Eldridge, M. and Grinter, B. (2001). Studying Text Messaging In Teenagers. Conference of Human Factors in Computing Systems (CHI) Workshop: Mobile Communications: Understanding Users, Adoption & Design

Farmer, R. (2003). Instant messaging - Collaborators Tool or Educator's Nightmare! In Proceedings of the Web-Based Learning Conference 2003 University of New Brunswick http://naweb.unb.ca/proceedings/2003/PaperFarmer.html.

Greenfield, P. M. and Subrahmanyam, K. (2003). Online discourse in a teen chatroom: New codes and new modes of coherence in a visual medium, *Applied Developmental Psychology*, volume 24, pp. 713-738.

Grinter, R. E., and Palen, L. (2002). Instant Messaging in Teen Life. Proceedings from Conference on Computer Supported Cooperative Work 2002. New York: ACM Publishers.

Herbsleb, J.D., et al. (2002). Introducing Instant Messaging and Chat in the Workplace. Conference of Human Factors in Computing Systems (CHI), Minneapolis, MN. pp 171-178. http://www.crew.umich.edu/Technical%20reports/Herbsleb_Atkins_Boyer_Handel_Finholt_ Introducing_instant_messaging_12_10_01.pdf

Isaacs, E., Walendowski, A., Whittaker, S., Schiano, D.J. and Kamm, C. (2002) *The Character, Functions, and Styles of Instant Messaging in the Workplace*. In Proc. ACM CSCW Conf., ACM, New York, 11-20. http://www.izix.com/pro/lightweight/images/IM.pdf

Lenhart, A., Rainie, L. and Lewis, O. (2001). Teenage life online: The rise of the instant-message generation and the Internet's impact on friendships and family relationships, *Pew Internet & American Life Project*, Washington, D.C.

Livingstone, S. and Bober, M. (2003). UK Children Go Online: Listening to Young People's Experience. Available online at http://personal.lse.ac.uk/bober/UKChildrenGoOnlineReport1.pdf

Nardi, B. Whittaker, S. and Bradner, E (2000). Interaction and outeraction: instant messaging in action. *Proceedings of the 2000 ACM conference on Computer supported cooperative work* ACM:USA pp.79-88 http://www.research.att.com/~stevew/outeraction_cscw2000.pdf

Oikarinen, J. (n.d.) IRC history. Available online at http://www.irc.org/history_docs/jarkko.html [accessed 9 Nov 2004].

Peris, R., Gimeno, M. A., Pinazo, D., Ortet, G., Carrero, V., Sanchiz, M. & Ibanez, I. (2002). Online chat rooms: Virtual spaces of interaction for socially oriented people. CyberPsychology & Behavior, 5, 43-51.

Schiano, D. J. et al. (2002). Teen Use of Messaging Media. Human Factors in Computing Systems: CHI 2002. Extended Abstracts. NY: ACM. http://hci.stanford.edu/cs377/nardi-schiano/CHI2002.Schiano.pdf

Scott, J. (2000) Social network analysis: a handbook. London: Sage

Smith, N. (2002). Teaching as Coaching, Helping Students Learn in a Technological World, EDUCAUSE Review, May/June 2002. http://www.educause.edu/ir/library/pdf/erm0233.pdf

Sulake Labs (n.d.) Habbo Hotel. http://www.habbohotel.co.uk/habbo/en/

Suler, J. (1997). From ASCII to holodecks: Cyberpsychology of an online multimedia community. *The Psychology of Cyberspace*, http://www.rider.edu/suler/psycyber/palsumary.html

Suler, J. (1998). Adolescents in Cyberspace: The Good, the Bad, and the Ugly. *The Psychology of Cyberspace*, http://www.rider.edu/suler/psycyber/adoles.html

Suler, J. (2000). Bringing Online and Offline Living Together: The Integration Principle. *The Psychology of Cyberspace*, http://www.rider.edu/suler/psycyber/integrate.html

Suler, J. (2004). The Final Showdown Between In-Person and Cyberspace Relationships. *The Psychology of Cyberspace*, www.rider.edu/suler/psycyber/basicfeat.html (article orig. pub. 1997).

Talamo, A., & Ligorio, B. (2001). Strategic identities in cyberspace. *CyberPsychology & Behavior, 4*(1), 109-122.

Thurlow, C. (2003). Generation Txt? The sociolinguistics of young people's text-messaging. *Discourse Analysis Online*. http://www.shu.ac.uk/ daol/articles/v1/n1/a3/thurlow2002003-paper.html

Williams, P. (1999). The net generation: The experiences, attitudes and behavior of children using the Internet for their own purposes. Aslib Proceedings, 51(9), 315-322.

Zakon, R. H. (2004) Hobbes' Internet timeline v7.0. Available online at http://www.zakon.org/robert/internet/timeline/ [accessed 9 Nov 2004]

BRIDGET JONES' IPOD

Relating Macro and Micro Theories of User Experience Through Pastiche Scenarios

Mark Blythe and Peter Wright

Department of Computer Science, University of York, U.K.

Abstract: This paper draws on macro and micro theories of user experience in order to focus on the use of Apple's iPod. It begins by outlining macro theories from Cultural Studies on the process of product articulation illustrated with reference to the Sony Walkman. The paper then turns to more recent theories of user experience generated in HCI over the past five years that offer micro accounts of experience and the ways in which a product interaction is shaped by particular people in particular contexts. These theoretical approaches are brought together in an account of the Apple iPod using pastiche scenarios that draw on popular fiction as a resource for reflecting on design. The account focuses on broad developments in the recording industry and also instances of individual iPod use

Key words: User experience, ubiquitous computing, wearable technology, acceptance and use of new technologies.

1. MACRO AND MICRO THEORIES OF USER EXPERIENCE

There have been many studies of technology in use carried out in Cultural Studies over the past twenty years and the theoretical accounts that emerged from them are often neglected in Human Computer Interaction (HCI) although they are often relevant. When the Sony walkman first appeared it caused much debate about the privatisation of public space: the user could effectively isolate themselves from their environment to create their own experience of it; problems of personal stereo etiquette such as volume and appropriate location caused, and to an extent continue to cause, concerns about isolation, alienation, manners, home taping and copyright infringement. Stuart Hall's *Doing Cultural Studies* traces the story of the walkman's production from original designs, to modifications following market research, to its adoption and integration in everyday life. Hall described "a circuit of culture" in which there were complex interrelationships between production, consumption, regulation, representation and identification (Hall and Du Guy 1997). The initial designs for the walkman for instance, envisaged two users listening to the device because it was believed that people would not want to listen to music by themselves. When it was discovered in early user tests that this is precisely what people wanted to do the design was modified. How the product was consumed then influenced later production. The study also traced, the development of the products name. This had to be "cool", early suggestions included "soundabout, stowaway" and "freestyle" (Ibid). Marketing campaigns were analysed to account for the process of representation in the media and the ways that people identified with the product, there was a concern then with what users thought the product said about them and how these messages were constructed. In order to account for the process of regulation the study focused on the recording industry's "home taping kills music" campaign in the nineteen seventies and eighties. This circuit of culture model can be thought of as a macro account of the ways in which cultural artefacts come to be embedded in everyday life. McCarthy and Wright (2004) in their analysis of technology as experience have argued that cultural analysis, valuable though it is to HCI has a tendency to downplay the role and diversity of individual experience.

In HCI there have been several attempts to account for user experience at the level of the individual (Ford and Forlizzi 2000, Davis 2004, McCarthy and Wright 2004). Wright and McCarthy argue that individual and cultural should be seen not as alternative perspectives on experience but rather as deeply inter-penetrating aspects of a more relational account of how we make sense of technology. They argue that the experience of technology is co-constructed by the designer, the user and the artefact with the user bringing values and meanings based on their own personal history and experience. Their relational or 'dialogical' approach suggests that our analytical methods should not force us to talk in terms of culture or individual experience but rather in terms of the individual-in-culture.

This paper attempts to combine the macro and micro approaches to user experience by reflecting on Apple's iPod. The next section presents a macro account focusing on the development of the recording industry. The section after that focuses on micro theories and considers the experience of use in pastiche scenarios.

2. THE RECORDING INDUSTRY: A HISTORICAL PERSPECTIVE

It has only been possible for musicians to make money from recording performances for a relatively short period of time. Although technologically possible since Emerson it was not until the mid nineteen twenties that gramophones were aimed at a mass market (Norman 1998). As the recording industry developed the sale of classical music made enough money for the particular companies to grow but few individual musicians could refer to the sale of their recordings as primary sources of income. It was not until the nineteen fifties that recordings of particular performances were bought by sufficient numbers of people to make a primary source of income for musicians. Up until around sixty years ago then, musicians made money by performing, not by recording their performances and selling that recording. The Beatles were the first group to become recording rather than performing artists. From Sergeant Pepper onwards the group stopped touring and the albums became the performances (MacDonald 1995).

When the Beatles launched "The Beatles" commonly known as "the white album" it was principally available on vinyl LP. As cassettes became more popular the situation became slightly more complicated and we witnessed the first great panic to seize the recording industry. Home taping, consumers were assured, was killing music. Although Beatles fans might have thought that they had bought the white album, the recording industry's position was that they had in fact merely bought a piece of circular plastic with grooves cut into it. The recording industry still owned the music and if the music became available in another format then consumers would have to buy that as well. And following the development of the compact disc this is exactly what many consumers did. They bought for a second time the albums they had previously owned in the LP format. This was, of course, a very profitable situation for the recording industry which made as much if not more from its back catalogue during the 1980s than it did from new recordings. So it was in *Men In Black* that Tommy Lee Jones could show Will Smith a newer, tinier music disc and tell him that this was going to replace CDs "Now" he remarks "I'll have to buy the white album again"

But the new format was not a newer smaller CD as the makers of that film imagined. The mini-disc never really took off as a replacement for CDs. The new format was not a tangible object it was a computer file. Marsall McLuhan famously argued that the medium is the message. The results of industrial automation were job losses regardless of the kind of machine in question (Mcluhan in Munns and Rajan eds 1995). In this sense the iPod is a fundamentally subversive technology. The final layer of its justly celebrated packaging is a cellophane wrapper bearing the inscription

"Don't steal music" in four languages. However, when it is connected to a MAC it immediately copies every song from the iTunes library into its memory without any reference to whether the music has been paid for or not: it is essentially amoral despite the manufactures admonitions. Increasingly music users are engaging in moral debates in order to justify and validate their experiences of technology. The British government have launched a music manifesto which will teach children that file sharing is wrong (Lettice 2004). However the interpretation of file sharing as theft can be and is contested (Woodworth 2004). Some music consumers then are beginning to think that they have bought the white album enough now.

Although mix tapes have long been produced by music fans the iPod allows for immediate re-organizations of material and for random elements to be incorporated. The iPod, unlike the walkman, introduces such elements to the listening experience as randomness. Someone listening to the white album on an iPod set to play tracks from it randomly will have a very different experience to someone listening to the album as it was made to be played: in a linear sequence of around twenty minutes per side. Indeed the Beatles spent a long time arguing about the sequence of songs on the white album and great care was taken to see that they flowed into one another (MacDonald 1995). With an iPod the user has far greater control, the user can literally co-construct the album. The technology liberates the user not just from the recording industry but also from the artist and the ways that they envisaged the experience of their music. Of course the music industry and certain artists within it (notably Metallica) have resorted to litigation in order to protect the existing models of music consumption – themselves accidents of technology. Legal means of file sharing such as the iTunes online shop and the iTunes network sharing facility are possible new models of music distribution and consumption but clearly we are still witnessing the process of articulation.

3. THE INDIVIDUAL IN CULTURE: PASTICHE SCENARIOS

A historical perspective on the development of the technology indicates that the use of any new media will be contested. This macro level account of the experience of technology has little to say about individuals with particular histories experiencing technology. Michael Bull has identified patterns of personal stereo use such as creating personal space, blocking intrusive sounds and images and disengaging from particular social situations (Bull 2000). These insights were drawn from interviews with personal stereo and iPod users but user studies are not the only way of reflecting on technology in use. Scenarios have long been used in HCI to reflect on use where either time, resources or access do not permit extensive user studies. A scenario, as the term is usually used in HCI, is a short narrative illustrating how a user might interact with some device or service providing sufficient detail to suggest design implications (Carroll 1997).They have been used to summarise the results of semi-formal user-centred requirements analysis (Monk 1998) and to summarise findings from field studies and other observations as the input to creating a design rationale (Carroll and Rosson 2003). They may also be useful in considering the impact of new technologies as they are adopted or articulated. Cooper (1999) argued that scenarios are only effective when used with personae. He argued that there are no typical users, no one, for example has two point three children. He suggested the construction of personae with names, ages, jobs and details about how they lived. Lene Nielsen has argued that users in scenarios are often mere functionaries that illustrate the workings of the product being described. She suggests drawing on traditions of European character driven film to generate more vivid and compelling characterizations.

Similarly, Wright and McCarthy (2004) have argued that the novel and in particular, the character-based novel is potentially a valuable tool for analyzing user experience since it provides a lens on the emotional-volitional nature of human agency. Pastiche is a style of writing that borrows from other texts and can be used as a method for creating rich personae in scenarios who represent individuals in culture. Pastiche scenarios then draw on existing narratives in order to create richer and more resonant descriptions of users and technologies (Blythe 2004).

The following scenario pastiches Bridget Jones' diary:

Guh! Almost missed train. Got on in nick of time but had no choice of seat. Just one left next to quite nice looking bloke but opposite nasty looking youth in baseball cap. Said youth plugged in v. loud headphones almost immediately so had no choice but to play nice new iPod. Took this out below table with some care. Wanted nice chap to see it (advertise self as successful young professional) but not youth in cap (must not see self as very much worth mugging.)

The identification with the iPod (self as glamorous young professional) is counterpointed by her fear that it makes her a target for crime (mugging by said youth). This anxiety can be thought of as one of the problems of what Reed calls "losable" technology as a counterpoint to the current term "wearable".(Reed submitted). The iPod has important symbolic functions then which are multiple and mutable.

Listened to Dear Prudence. V. good song. Reminds me of first year at uni and Marcus. Nice bloke and Beatles fanatic. Apparently John Lennon wrote said song for Mia Farrow's sister. They were in India being all spiritual and following the Maharajah or similar. This Prudence went off deep end and meditated for about three days straight in manner of scary hermit. Think of it - Paul McCartney and John Lennon outside strumming guitars and probably in need of severe seeing to after all the chanting and what not, and Prudence decides she'd rather sit in dark hut. Inappropriate lyrics current weather-wise. "The sun is out, the sky is blue, it's beautiful and so are you" I think not John, I think you'll find that the clouds are low, the sky is grey and the awful youth in front of me is picking his nose. Hippies. Mean well I suppose but it just isn't practical. All you need is love. Pah. What about chocolate and fags? Not to mention a nice Chardonnay.

Her memories about other encounters with the white album, the stories she knows about the Beatles; her images of India; the atmosphere of the music; the lyrics of the song; her imagination and the technology delivering the sounds combine to make a unique experience. This experience is both an internal and an external event.

"Look around around around around round" Beatles suggest. I do and the landscape doesn't seem so bleak. For a moment I feel all sixties. Yes, everything is beautiful, even youth in front of me. And of course he's not a mugger just because he likes baseball caps. Peace, love, yes, why not? Then abrupt stop. Forgot to charge wretched iPod so spent rest of journey listening to youth's tinny headphone sound and trying to work out why he was listening to swarm of bees humming song by Metallica. Then it occurred that youth was listening at least to a CD, not stolen songs like self on one woman crime wave.

Although it is commonly argued that personal stereos are alienating, turning what was once a communal activity (listening to music) into a personal one and making public spaces private, the scenario illustrates the ways in which experiences of technology can be as intense and moving as any experience of art.

> "at its height it [experience] signifies complete interpenetration of self and the world of objects and events" (Dewey 1934: 19)

The user brings their own past, constantly changing expectations and reflections to bear on an experience. The user's personal history, knowledge and values then play a large part in the experience and the technology allows for the expression of a range of values, contrary to those of the music industry and the law as it stands. ITunes are still negotiating for the Beatles back catalogue. The only way currently that Bridget could have got Dear Prudence onto her machine is by copying a CD she owned or one that she did not own.

> *Is this on? Not sure. Supposing had bought white album instead of lending it from Shaz and copying it onto computer. Money would go straight to Yoko Ono and whinging Julian Lennon. Seem to remember John Lennon saying he did not want his kids turning into new aristocracy or similar. Quite right. Must save poor Sean from terrible fate of perpetual payment. Shouldn't buy Beatles again on principle. Intellectual property is theft. Hmm, quite a ring to it that. Must write some sort of article on same.*

The experience of the iPod is made up not just of stories about the songs creation, her own memories and the song itself, but also of her moral and political attitudes to the technology: the experience of the iPod then, is not only social, cultural and aesthetic but also moral and political.

It may be argued that Bridget Jones is a stereotype. Viewers of the recent film sequel would be justified in thinking so. But the character in the book is much more than that, the text could not have resonated at the level it did were it otherwise. Bridget Jones may be a caricature but she is certainly more vivid than the kind of stereotype a conventional scenario might evoke and so a useful device for thinking about the micro levels of experience that recent HCI theories articulate. Patrick Jordan has argued that method acting can be a useful way for researching groups (criminals for example) who cannot be reached by more conventional methods (Blythe 2004). Pastiche scenarios might also be used in this way. Imagine, for instance, that the youth in the baseball cap opposite Bridget Jones in the last set of scenarios is one of Irvine Welsh's Trainspotting characters, Renton. In the following scenarios, Renton sits opposite Bridget having quite a different experience of the same environment.

> *This Velvet Underground track is making me wish I had some skag. It is impossible to maintain an anti-drug position and listen to great music. If Paul McCartney had done more hardcore drugs he would have composed less shite like the frog's chorus, the result, evidently of wine and marajuana. "When I'm rushing on my run And I feel like Jesus son" Yes, I could do with some of that. But no I tell myself I'm off that shite for good now. When this train meets its destination Mark Renton arrives a new man, a man on a diet high in moral fibre. Aye, maybe just one more hit though eh? And that woman has in her possession one skag voucher of the highest order, an iPod. White earphones, dead giveaway. Worth*

about three hundred quid retail, on the open market, more than enough for one last hit. And she's only popped it into her right hand jacket pocket, just near the aisle, perfect.

Here then is a perhaps infrequently recognised affordance of an iPod, not only is it a machine for playing music it is also a means of raising cash. The portability so attractive to users is also a useful feature for thieves. Mobile phone theft is one of the highest growing crimes in the UK, as palm computers, PDAs and iPods increase in popularity so too will, if not actual crime rates, then fear of crime rates. These items are what Reed (submitted) would call "stealables".

Some gadge in a suit is saying something to me, I remove an earpiece and raise a laconic eyebrow in the style of the young Sean Connery
- Excuse me I think you're in my seat.
- I don't think so, I reply with as straight a face as possible. - I reserved this seat in good faith.
That's strange, can I have a look at your ticket?
- Certainly not, I know my rights.

Everyone around us is pretending not to listen. The woman is shitting herself, all of them just wishing I'd go away.

I stall him and the tension builds. Eventually I ask to see his ticket.
- I'm terribly sorry, clearly there's been some terrible misunderstanding. Fucking GNER or whatever you call the privatised shite eh?
I flash round a friendly smile but everyone looks away, including the target. I stumble against her and slip the skag voucher away unnoticed.
Sorry madam, I do apologise.
I laugh and she looks terrified. We're almost at the station and one last hit. "Enjoy." Thank you Mr iPod, don't mind if I do.

Ironically, just as Bridget Jones is debating the morality of copying music (or stealing music if the recording industries' construal is accepted) Renton steals her iPod. The use of pastiche prevents the scenario from becoming one of an ideal use situation, Bridget's hapless and disorganised character suggesting that she would not have charged her iPod. Renton's addiction suggesting the affordances of the iPod in terms of its stealability. The iPod slogan "enjoy" is re-situated by the character and its meanings are subverted.

I had to buy a charger so as to make it a better buy. The woman's music collection represented a shocking state of affairs. The only decent album on it was by the Beatles. Her top 25 playlist had All by myself at the top!

All by Myself! I'm no fucking surprised love listening tae crap like that. That track alone knocked at least twenty quid off its value. Trying to explain the technological possibilities of the machine to drug dealers was an exercise in futility. I tried to tell them that all they had to do was steal a MAC and they'd be sorted but they weren't biting. More skag inspired musicians required in that music collection. The white album, case in point. Lennon off his tits on skag writes Happiness is a Warm Gun. McCartney contributes music hall shite like Rocky Racoon. I need a fix cos I'm going down.

When an iPod is stolen so too is an entire music collection. This is likely to have far more value to the person who has lost it than the thief. The iPod, containing playlists of the most frequently played songs is highly personalised and its loss might be felt quite profoundly. Bridget and Renton's opposing notions of what constitutes good music and their widely different readings of the white album indicate the particularity of experience even with regard to widely popular cultural artefacts. Marc Davis (2003) drawing on reception theory points out that when we read cultural texts (both written and in other media) we bring as much to the text as the text brings to us in order to form a gestalt – something greater than the sum of its parts. The white album means very different things to Bridget and Renton and so too does the iPod.

These scenarios defamiliarise the familiar and this in itself can be a useful resource for design (Blythe, Bell and Sengers in press). The scenarios defamiliarise the use situation and perhaps also the genre of writing – the scenario. They also illustrate some concerns which may be addressed by design. For example one design implication is that while the distinctive iPod design is valued as a status symbol it may be useful if that appearance could be modified at times in order to relieve the fear of crime if not prevent actual crime itself. Similarly iPods might be given more prominent security settings such as passwords of the kind available on many mobile phones which might make them less usable to a casual thief. Alternatively they could be made cheaper so as to be less appealing to thieves. These implications are merely examples of how such approaches might be of use to design. Because the source material is rich the scenarios are ambiguous, other readings and so design implications may be latent in these scenarios.

4. DISCUSSION AND CONCLUSION

In *Resistance Through Rituals* Stuart Hall (1982) reflected on the youth culture of the sixties and seventies. While the peace movement of sixties hippies and the seeming outright rejection of mainstream culture embodied in the punk movement of the seventies were deeply political and in some senses revolutionary, Hall points out that this same resistance was almost immediately absorbed, repackaged and sold back to the sub groups that originated it. Thus fashion shops began selling the safety pins that so alarmed the parents of the punks that put them through their ears and noses. Current popular culture is frequently characterised as apolitical and apathetic. Cultural commentators refer nostalgically to the revolutionary spirit of the sixties counter culture, the rebellion and energy of the seventies punk movement. The current music scene and the artists dominating it are seen as bland, corporate representatives. But the music fans who are engaged in file sharing and building trans-national communities of music sharers can be seen as far more revolutionary than their predecessors. They may also be more effective.

File sharing and CD copying have reduced very powerful international corporations to a state of free fall panic where they are attempting to litigate their way out of the problem (Woodworth 2004). Assuming that they will not be able to criminalise a very large part of the population this may be one form of pop cultural resistance that cannot be incorporated and commoditized. The use of the iPod then is deeply political. In an age where capital presses ever further in to our every day lives and even has the temerity to trademark DNA, here is a bastion of resistance to copyright culture. But iPods and other music file playing devices are in themselves highly stealable. As the gap widens between the technology haves and have nots crime and the fear of crime are likely to become increasingly important problems for designers to address not just for music producers but music consumers.

Emerson's gramophone offered users new ways of experiencing music. For the first time it was possible to listen to music in the absence of musicians. Music could be enjoyed at home. Recorded music is now an integral part of our lives accompanying many of our daily activities. Digital media, like the earliest recording technologies, offers radically new forms of experience. In time the music industry will adapt, although the process of articulation may be a slow and painful one.

Geertz (1986) suggested that cultural analyses can appear to float several feet above their human ground, but likewise, psychological analyses can seem to be buried several feet beneath. A dialogical approach to user experience aims to analyse the individual-in-culture and to explore the interpenetration of culture and individual in understanding how people make sense of technology in their lives. A commitment to a dialogical view of individual-in-culture is a commitment to ways of analysis that do not treat these as separate components. Bakhtin argued that the character-based novel is the most sophisticated tool available for seeing, valuing and expressing human experience in all its richness (Morson and Emerson 1990). By bringing some of the techniques of the novelistic genre into scenario-based design, pastiche scenarios offer a way in which we can get a glimpse of the richness and complexity of the individual-in-culture and incorporate macro and micro accounts of user experience.

REFERENCES

Bell, G., Blythe M., and Sengers P. (in press) Making By Making Strange: Defamiliarisation and the design of domestic technologies. TOCHI special issue on HCI and social issues

Blythe M. (2004) Pastiche Scenarios. *Interactions* magazine

Blythe M. (2004) Interview with Patrick Jordan. *Interactions* magazine

Bull M. (2000) Sounding Out the City: personal stereos and the management of everyday Life. Oxford Berg.

Carroll J. M. (1997) Scenario-based design. In: Helander M., Landauer T.K., and Prabhu P., Editors, *Handbook of Human-Computer Interaction, Second Completely Revised Edition,* Amsterdam, Elsevier.

Cooper A. (1999). *The Lunatics Are Running The Asylum: Why High-Tech Products Drive Us Crazy and How to Restore The Sanity* (Sams, Indianapolis)

Davis M. (2003) Theoretical Foundations for Experiential Systems Design. ETP 03 2003 Berkeley, USA ACM 45-52

Dewey J. (1934) Art as Experience Minton, Balch and Company, New York.

Geertz, C. (1986) Making experiences, authoring selves. In E.M. Bruner, E.M. and V. Turner, V. (eds.). *The Anthropology of Experience*, p.373-380. Urbana: University of Illinois Press.

Gramsci A. (1971) Selections from the Prison Notebooks. Lawrence and Wishort. London

Fielding H. (1999) Bridget Jones' Diary. Penguin, London.

Forlizzi J. and Ford S., 2000, The Building Blocks of Experience: AN Early Framework for Interaction Designers. DIS 20000 (Brooklyn, New York. DIS) 419–423

Hall S. and Jefferson T. (1982) *Resistance Through Rituals: Youth Subcultures in Post-War Britain* Hutchinson and Co. London

Hall S. Du Gay P. (1997) Doing Cultural Studies: Story of the Sony Walkman (Culture, Media and Identities). Sage Publications

Lettice J. "Stealing Songs is Wrong" Lessons Head for UK Schools. The Register 5. 8.04

McDonald I. (1995): Revolution in the Head: The Beatles' Records and the Sixties. Henry Holt and Company

McLuhan M. (1995) The Medium is the Message in Munns and Rajan (eds) A Cultural Studies Reader: History, Theory and Practice. Longman Group Limited

Morson, G.S. and Emerson, C. (1990). Mikhail Bakhtin: Creation of a Prosaics. Stanford, California: Stanford University Press.

Nielsen, L. (2002) From User to Character: An investigation into User-descriptions in Scenarios in the *Proceedings of DIS2002* , The British Museum, London 25-38 June

Norman, D., 1998, *The Invisible Computer: Why Good Products Can Fail, the Personal Computer is so Complex and Information Appliances are the Solution.* (The MIT Press Cambridge, Massachusetts, London England).

McCarthy J. and Wright P. (2004) Technology as Experience. MIT press

Reed D. (submitted to HOIT 2005) Learning From Losables: An Exercise in Critical Reflection.

Woodworth G. M. (2004) Hackers, Users, and Suits: Napster and Representations of Identity. *Popular Music and Society* Vol 27 Number 2 pp161-184

Welsh I. (1996) Trainspotting. Minerva, London

Wright P and McCarthy J. (2004) The Value of the Novel in Designing for Experience in A Pirhonen, C. Roast P Saariluoma, H Isom (eds) Future Interaction Design

THE IMPACT OF THE SOCIAL ON THE HOME COMPUTER

Lynne Baillie
Telecommunication Research Center, Vienna, Austria.

Abstract: Why do some technologies successfully assimilate into the home and others fail? The physical layout could be one factor but another more important factor could be the social aspects of the home space. Also, once a technology enters the home does it impact on the social life of that home? Some manufacturers in the past have adapted their devices so that they assimilate more clearly into the home, however, until quite recently this was something that was not done by manufacturers of PCs. The impact that this design decision has had on the acceptance of the home computer is the topic of this paper.

Key words: domestic technology, design decisions, acceptance.

1. INTRODUCTION

This paper investigates the findings of studies on the use of the PC in the home, this is important as the background research provides a rationale as to why the PC for a long time was not assimilated into the social aspects of the home. The first section reports on some early studies of the adoption and use of computers in the home and then explores whether the PC has ever really become a 'home' computer in the broadest sense. The importance the advent of the Internet has had on the home computer is then discussed. The paper concludes with a short discussion of what implications these findings may have for the new more entertainment orientated PC's currently being proffered for the home.

2. THE HOME COMPUTER AND THE INTERNET

2.1 A Brief History of the Home Computer

Home computers were sold to people in the 1980s as a way of working at home, to help educate children, for word processing, and generally equipping yourself and your children for the future. However, enthusiasm for this idea levelled off. Researchers found that some of the reasons for this were that there was not a functional need for a computer in the home (except for a small minority of people), the word-processing function could be equally well performed by an expensive typewriter or dedicated word processor, and the video games function could be fulfilled more cheaply by a home video game machine (Caron, 1985). Because of this the families that had bought computers in the 1980s failed to trade-up to more sophisticated models, when they became available on the market e.g. Amstrad's PCW IBMs home computer, BBC Micro, Sinclair Spectrum and Dragon 32. Murdock, Hartmann and Gray (1992) thought that early disillusionment had affected the views of householders. A quote from one of the householders they interviewed:

> 'I wanted it as a word processor, but of course its no good for that at all. I didn't appreciate it at the time. You can't get enough words on the screen, unless you get one that's about four or five times the price...I wanted it for a word processor and they said (the shop assistants), "Oh yes they can do this" and "Oh yes they do that", you know. And of course it does, but not satisfactorily for proper use.' (Murdock et al. 1992, p.148)

This quote clearly highlights why people were disillusioned with the early computers they bought, as the PC quite singularly failed to live up to the marketing hype and was not robust enough to perform the tasks that were required of it e.g. word processing. Another reason for failure was highlighted by Silverstone et al (1992). They commented that computers had failed to find a place within the moral economy of the household, specifically in terms of its incorporation into the routines of daily life. This failure is discussed in the next section.

2.2 The 'Home' Computer

One of the comments made about PCs in the home is that they have failed to be completely assimilated into that environment (Venkatesh, 1996; Silverstone et al., 1992). The look of a PC could be one reason why the PC bought for the home place has, until quite recently with the advent of iMacs, failed to be assimilated into the home as the PC in the home-place has looked no different from the PC in the workplace. PC manufacturers have not until quite recently acted in the same way as Singer (see Section 3) and try to redesign or remodel the PC so that it would be better assimilated into the home. The PC in the 1980's and 90's was still in the main a stand-alone unit that had come from an industrial/business background, and unlike other technologies in the home, had not been specifically built to carry out a task in the home e.g. cleaning floors. Some researchers suggested in the 1980s that the reason for this is that the PC is too complex a machine for the home. Rubin commented that unlike other technologies in the home:

> 'Learning to use a computer is much more like taking up a musical instrument than following instructions on how to use an electrical appliance, such as a toaster.' (Rubin, August, 1983).

However, research carried out at that time and since has suggested that people do not use their PCs in complex ways, for example, Caron (1985) found that people in the home only used their PC for word processing and games. Similarly in the 1990's Kraut (1996) found that people mainly used their PCs for games and email. Their research suggests that people use and have historically used their PCs in the home in quite straightforward and uncomplicated ways. Therefore, it cannot be assumed, that just because people are using a complex machine they are using it in a complex way. It could be concluded then that the poor assimilation of the PC into the home cannot completely be blamed on the complexity of the machine.

A possible alternative suggestion, as to why the PC has experienced poor assimilation, is that people in the home are unsure of where to place it. During their research, the team at CRITO (Center for Research on Information Technology and Organizations), found that in the 1980s the personal computer, although in the home, had not become a 'home computer' and had not been integrated into the home in the social context. They thought that there might be several possible reasons for this: people viewed their home computer as an extension of their work computer; software for home use was not well developed; no telecommunications connectivity; inapplicability and inappropriateness of the industrial/business model for the household; and a lack of computer skills amongst the general public. Subsequently, the adoption of PCs grew quickly in the late 1990s, according to Venkatesh because there was a new generation of young adults who grew up with PCs, more manufacturers targeted their software towards the home market, and most importantly PCs had acquired communication capabilities. The importance of these new communication capabilities are of critical importance when one comes to consider the PC in the home as a social facilitator.

2.3 The Impact of the Internet on Home PCs

The impact of the Internet on home computers cannot be underestimated. However, Kraut (1996) thought that this marketing of these new internet capabilities was flawed. He compares this with how the telephone was at first marketed for business use and then later in the home:

> 'Manufacturers of the telephone did not foresee that many different types of people would like to talk on the phone for its own sake and not to accomplish a task, but because they enjoyed talking to other people.' (Kraut, 1996, p34).

He comments that how banks, phone companies, and computer companies insisted on selling computers and online services to consumers for telecommuting, home banking, and home shopping, seemed to echo the early misunderstanding of the use of the telephone. The main use of the telephone was in fact for personal communications, this has also been found to be the main use of the Internet in the home (Kraut, 1996). Venkatesh (1996) made a similar finding commenting that business people stress the mass media features of the Internet and its ability to broadcast information, entertainment, and advertisements to a large population of consumers. In contrast, people at home value the Internet for more personal uses to maintain social relationships and to participate in their local communities. However, even with these new telecommunication capabilities the home PC still failed to permeate homes in the same way as television and the telephone. In the next section I discuss how it could be another issue that of the look and design of the PC itself could have had a major impact on the social acceptance of the PC in the home.

3. RE-DESIGNING FOR SOCIAL ACCEPTANCE

The sewing machine is an interesting case of when a manufacturer completely changed the look of the machine so that it would be accepted in the home. Sewing machine manufacturers in the mid 1800's were faced with a massive drop in industrial sales. A possible market that had not been sold to before was the home. It became an economic necessity, therefore, to turn the sewing machine into a domestic appliance, but this was easier said than done. Initial attempts were not too successful, mainly because the industrial connotations had not been removed. The manufacturers realised that unless they created the impression that the place for a sewing machine was in the home, the market was never going to accept the machine. The manufacturers did not try to redesign the sewing machine but merely place it in their marketing literature in the setting of a parlour. These adverts, however, did not convince people to buy. Because as Forty (1986) commented having these sewing machines in your home would be like having a machining tool in your living-room. One manufacturer (Singer), however, decided to go a step further than the others and design the machine to fit into the home setting. They did this by making the machine smaller in size, lighter in weight, and giving it a more elegant design. This led to massive sales and to the acceptance of the sewing machine in the home.

It could be concluded from this that computer manufacturers in the 1980s and 1990s failed to do what Singer did and redesign the PC to fit into the social space of the home. How some manufacturers are aiming to change that is discussed in the next section.

4. THE FUTURE FOR THE PC AND OTHER TECHNOLOGIES IN THE HOME

Mackay carried out a study of people's acquisition of the following devices: telephone, television, videocassette recorder (VCR), home computer, computer games console, and compact disc (CD) player. He was interested in these technologies because he thought that they were becoming more and more connected and inextricably linked. He concluded from his studies that there is a growing demand for interactivity between PCs, games consoles, and CD-Players. Recently, the mobile telephone has been added to this list. The scale and significance of this shift should not be underestimated he said because in recent years the video game has superseded popular music as the main form of entertainment:

> 'Far from being a marginal arena, the home is a key location for our experience of IT, and for the success (or otherwise) of specific IT products. Domestic leisure as we now know it is inconceivable without an extensive range of IT devices. Rather than being marginal, or 'not real IT', these devices are absolutely central to IT markets, IT developments and our daily experience of IT.' (Mackay, 1995, p267).

Ted Turner - CEO of cable news network- made the following comment about interactive television:

> 'Every single interactive TV experiment has failed. Most people want to sit back and watch- interacting is hard work'. (Ted Turner, 1994, p3).

So do we want to interact in new ways with our devices? It would seem in the case of interactive television, that we do not, in a report by Towler (2001) for the Independent Television Commission, it was reported that 80%

of the people with digital televisions said that they never or only rarely used any of the interactive services and that they had in fact bought the TVs because of the improvements in sound and picture quality. So we do not seem to want to interact more with our televisions and our PCs are being left to gather dust.

One technology firm has tackled this issue by marketing their entertainment PC as a television first: "The PC to replace your television" (Sony, 2005). It would seem that Sony are following the path forged by Singer's (see previous section) and re-designing the PC to fit more comfortably into the home space as Jun Koyama (2004) from Sony comments:

> "Our new PCs are for people who don't want to see PC as a beige box in the corner of their living room that performs a useful but essentially dull service. It is intended to sit centre stage, and act as an entertainment system."

This approach though could be problematic in the home as it means that what were three-four separate devices are now contained in one unit. Researchers have criticised this approach as they say that these new systems containing one large display ignores the fact that physical materials and technologies are currently distributed throughout the home to suit that particular household's lifestyle (Graves-Petersen, 2004). Furthermore people are wary of so much technology being contained within one device, as this would mean that if the system broke down they would not only be without their television but also their computer, games console and sound system. Another hurdle to overcome in the immediate future is the lack of standard communications between devices from different manufacturers.

However, there is a demand for ever more linked up technology and interactivity and these technologies will be combining the power of a computer with the latest communication technology. At the moment, however, it would seem that the user is not being consulted on the design and functions of these newly minted technologies and may well be left as the outsider. An improved approach to the field of dreams (i.e. if we build it they will come) would be if manufacturers and designers tried to understand:

- What technologies people want to be able to connect together and when?
- The level of interactivity they want.
- How to provide these services via the device (perhaps through the infrastructure of the web).

5. CONCLUSION

At the moment mergers are taking place of several technology sectors e.g. computers, telecommunications, broadcast and cable television. To facilitate this merger and to make the new devices and services financially viable, manufacturers 'need' people to use these new interactive devices and services in their homes. This requires users to interact with their devices in new ways e.g. carry out their banking via their television, or ordering food on-line via their microwave. It could be strongly argued given the research outlined in this paper that the very fact that these devices are in peoples homes is not enough in of itself to guarantee that people will use them in the way that manufacturers want and crucially 'need' them to use them. This may then be the perfect time to study once again our relationship with the home PC and other home technologies as our relationship with these technologies is set to become more complex with the advent of digital television, home media centres, and hybrid telephones.

REFERENCES

Campbell, C. (1992) The Desires for the New: Its Nature and Social Location as Presented in Theories of Fashion and Modern Consumerism. In R. Silverstone., & E.Hirsch (Eds.) Consuming Technologies: Media and Information Spaces.(p.54-65).London: Routledge.

Caron, A.H. (1985) The Process of Microcomputers in the Home: Uses and Impacts. Montreal Universite de Montreal. Departmente de Communication. Unpublished Paper.

Corrigan. P (1997) The Sociology of Consumption. London: Sage

Csikszentmihalyi, M., & E. Rochberg-Halton (1981) The Meaning of Things: Domestic Symbols and the Self. New York: Cambridge University Press.

Forty, A. (1986) Objects of Desire: Design and Society Since 1750. London: Thames & Hudson.

Graves-Petersen, M. (2004) Remarkable Computing- the Challenge of Designing for the Home.

Johnson, S. (1997) Interface Culture: How New Technology Transforms the Way We Create and Communicate. New York: Basic Books.

Kraut, R. (1996). The Internet@ Home. ACM, 39 (12): 32-35.

Koyama, J (2004) http://www.sony-europe.com/PageView.do?site=odw_en_EU&page=PressReleaseDetail&section=en_EU_Press&pressrelease= 1094201879226

Miles, I., Cawson, A., & Haddon, L. (1992) The Shape of Things to Consume. In R. Silverstone., & E.Hirsch (Eds.) Consuming Technologies: Media and Information Spaces. (p. 66-83) London: Routledge.

Mackay, H. (1995) IT and the Home. In N. Heap., R. Thomas., G. Einon., R.Mason., & H.Mackay (Eds.) Information Technology and Society: A Reader. (p. 267-311) London: Sage.

Murdock, G., Hartmann, P., & P.Gray. (1992) Contextualising Home Computing: Resources and Practices. In R. Silverstone., & E. Hirsch (Eds.) Consuming Technologies: Media and Information in Domestic Spaces. (p.146-162) London: Routledge.

Norman, D. (1999) The Invisible Computer. London: MIT Press.
Silverstone, R., Hirsch, E., & Morley, D. (1992) Information and Communication Technologies and the Moral Economy of the Household. In R. Silverstone & E. Hirsch (Eds.) Consuming Technologies: Media and Information in Domestic Spaces. (p.15-30) London: Routledge.
Sony (2005): http://www.sonystyle.com
Towler, R (2001) The Public's View 2001. An ITC/BSC Research Publication. http://www.itc.org.uk/itc_publications/index.asp (last checked 19/07/2002).
Turner, T. (1994) What if they're right? The Economist. Feb 12, p3.
Venkatesh, A., & Vitalri, N. (1986) Computing Technology for the Home: Product Strategies for the Next Generation. Journal of Product Innovation and Management, 3(1), 171-186.
Venkatesh, A. (1996) Computers and other interactive technologies for the home. ACM, 39 (12), 47-54.

HOME NETWORKING

The role of mobile ad hoc network in future home networking environments

Humayun Bakht, Madjid Merabti, and Bob Askwith
School of Computing and Mathematical Sciences, Liverpool John Moores University, U.K.

Abstract: Mobile ad-hoc networks are the networks that can be formed with the mutual coordination of two or more mobile devices without the intervention of any fixed infrastructure. Our future homes and Industrial environments are likely to be based on the networks which are easy to deploy with minimal installation and maintenance cost. Mobile ad-hoc networks have strong potential to fulfil the requirements of future computing. Where, on one hand mobile ad-hoc networks are able to give us a networking environment where we can form network on our convenience without involving much complexities. The other side shows that there are number of issues in this area which require satisfactorily solutions before we will be able to use mobile ad-hoc networks at a wider scale i.e. home, business, education and other environments. Routing can be seen as one of the important aspect of any networking environment. It is an unresolved issue in mobile ad-hoc networks. Work is going on since over a decade to invent an efficient routing mechanism for mobile ad-hoc networks. Numbers of different techniques and algorithm have been proposed. However, most of these solutions have certain limitation which makes them unable to operate in all environments. Moreover, very less attention has been given to obtain analytical results of these algorithms in a real environment. Our research shows, that most of the existing issues in mobile ad-hoc networks are interrelated with the routing mechanism of mobile ad hoc networks. On the basis of carried out research, we have also concluded that most of the proposed protocols, if good in addressing some issues besides routing, they are not very impressive in handling others. As deployment of mobile ad hoc networks in future home environment depend very much on the invention of a successful routing solution. We have proposed a novel algorithm 'mobile ad-hoc on demand data delivery protocol (MAODDP)' as a routing solution for mobile ad hoc networks. One of the important features of MAODDP is the fast communication by establishing route and delivering data simultaneously at the same time.

MAODDP follows an intermediate approach in between tables driven and on-demand routing protocols and can be used to save bandwidth and battery life of the participating nodes of mobile ad hoc networks. In this paper, we will discuss in details about various aspects of mobile ad hoc on demand data delivery protocol and its importance in our future computing environment.

Keywords: Wireless, networks, protocols.

1. INTRODUCTION

Mobile ad-hoc network is one of the new additions in the family of wireless networks[1]. It is an autonomous systems of mobile nodes connected together forming an ad-hoc or temporary network in the absence of fixed infrastructure[2-4]. Operating in the absence of fixed supporting structure poses several different challenges to mobile ad-hoc networks[5]. Some of those challenges include routing, security, hidden terminal problems and bandwidth constraints[5, 6]

Routing[7, 8] is the mechanism of information exchange between any two hosts in a network; it is an important aspect to be seen. Routing is one of the challenging issues in mobile ad-hoc networks. Existing Internet protocols were designed to support routing for environment with supporting structure. Their performances therefore, are not very impressive on mobile ad-hoc network.[9] This fact results in the development of new routing strategies for mobile ad-hoc networks. In mobile ad-hoc network a route between a pair of mobile nodes may go through several other mobile nodes. These routes can change when hosts change locations. One traditional way of achieving routing in mobile ad-hoc network is to consider each host as a router and running some conventional routing protocol. This approached is not feasible for mobile ad-hoc network, which suffer with unpredictable and frequent topology changes. Good numbers of papers have been reported proposing various routing solutions[10-14]. One of the main weaknesses in most of the proposed solutions is the lack of consideration given to other routing related issues. Moreover, these protocols do-not perform well in all environments[9, 15, 16] .In an ad-hoc network, all nodes cooperates each other in order to establish and maintains routing in the network, forwarding packets for each other to allow communication between nodes not directly within wireless transmission range. Rather then using the periodic or background exchange of routing information, as it is common in most of the routing protocols.

The contribution of this paper is to highlight in detail about the possible use of mobile ad-hoc networks in the future home environment. In this paper, we have also included a detail description of mobile ad-hoc on demand data delivery protocol as a routing solution for mobile ad-hoc network. Rest of the paper has been organized as follows. Section 2 throws light on the possible implementation of mobile ad-hoc networks in future home networking environments. Section 3 details the brief description of mobile ad-hoc on demand data delivery protocol (MAODDP). Section 4 covers conclusion and future work while references are listed in section 5.

2. MOBILE AD HOC NETWORKS IN FUTURE HOME NETWORKING ENVIRONMENT

Recent advancement such as Bluetooth[17, 18], WIFI and ad-hoc sensors networks[19] have reviewed an old concept of ad-hoc or peer-to peer networks. Mobile ad-hoc networks are the networks of two or more mobile devices connected with each other without the intervention of any fixed infrastructure. Where on one hand mobile ad-hoc networks offer benefits which cannot be avail from other networks technologies of similar kinds[20]. On the other hand, there are still number of different issues awaiting suitable solutions before we can expect to see their deployment at a wider scale.

The focus of the current ongoing research in this area is to invent various techniques to support different network controls. One of the main aims of ongoing research is to ease end-users jobs by reducing installation and maintenance hassles. At present, setting up a wireless environment in a home environment not only costly but also there are several other issues that are involved. Not every one has technical knowledge to choose the best wireless product which can fulfill one's needs at a lower rate. Moreover, if one is not an expert in setting up network systems, chances arc the user's needs to communicate with technical peoples to resolve various problems which may arise with time. This issue not only causes extra cost but also create unnecessary burden on the end users. Likewise, very rarely we find people who are willing to learn different concepts of network terminologies. Setting up a wireless system at its present state requires end users to have at least basic networking knowledge in order to do minor changes on day to day basis whenever it requires. One final and important issue is the continuous updating of the installed system. It happens that the system hardware and the network supporting software could be outdated with time. Therefore, to achieve faster and better services, end users might require updating different parts of their system.

Therefore, there is a solid need of a network system which can easily be used by both technical and non technical users. The system can easily be available, easy to understand without involving too much complexity.

Mobile ad-hoc networks offer these features. Successful deployment of these networks could possibly create an environment where we can use various home objects such as washing machines, oven and cooker for networking purpose. Ad-hoc networks can easily be deployed and can be well suited in all environment i.e. home, Industrial, educational and banking. Another area that has attained a point of focus is how to connect an ad-hoc network with a larger network such as the Internet. Work is on its way to establish mechanisms that can possibly ease mobile ad -hoc networks connection with the larger networks. One final issue is the network security both for homes and business places. Ad-hoc sensor networks which are one of the implementation of mobile ad-hoc networks are currently in used to secure different places such as shopping malls, parking lots etc.

3. MOBILE AD-HOC ON DEMAND DATA DELIVERY PROTOCOL (MAODDP)

Based on the idea of faster and efficient data delivery, MAODDP established the route between the source and the potential destination on demand and deliver the data simultaneously at the same time. MAODDP provides loop-free routes through the use of sequence numbers associated to each route. In short, if A needs a route to B it broadcasts a ROUTE REQUEST and data delivery packet (RREQD). This packet contains the desire destination information the source node aware of and the packet to deliver. Each node that receives this packet, and does not have a route to B, rebroadcasts it. The node also keeps track of the number of hops the message has made, as well as remembering the source of the RREQD. If a node has the route to B it forwards the packet to B.

To achieve faster convergence in the network and thus higher mobility, a ROUTE ERROR message can be broadcasted on to the network in the case a link breakage occurs. Hosts that receive the error message remove the route and re-broadcast the error messages to all nodes with information added about new unreachable destinations.

3.1 Overview

When a route to a new destination is needed, the node uses a broadcast RREQD packet to find a route to the intended destination. A route can be determined when the RREQD reaches either the destination itself, or an intermediate node with a 'fresh enough' route to the destination. A 'fresh enough' route is an unexpired route entry for the destination whose associated sequence number is at least as great as that contained in the RREQD.

Once the data delivery accomplished successfully, an acknowledge message (ACK) is send back to the source of the message. Each node receiving the RREQD store a route back to the source of the request, so that the acknowledge message (ACK) can be unicast from the destination along the recorded path to that source, or likewise from any intermediate node that is able to satisfy the request. Nodes monitor the link status of next hops in active routes. When a link break in an active route is detected, a RERR message is used to notify other nodes that the loss of that link has occurred. The RERR message indicates which destinations are now unreachable due to the loss of the link. In order to enable this reporting mechanism, each node keeps a ``precursor list", containing the IP address for each its neighbors that are likely to use it as a next hop towards the destination which is now unreachable.

Each node operating on MAODDP maintains a route table management. Route table information must be kept even for ephemeral routes, such as are created to temporarily store reverse paths towards source nodes broadcasting RREQDs. MAODDP uses the following fields with each route table entry:

- Destination IP Address
- Destination Sequence Number
- Hop Count (number of hops needed to reach destination)
- Last Hop Count
- Next Hop
- List of Precursors
- Lifetime (expiration or deletion time of the route)
- Routing Flags

Managing the sequence number is crucial to avoiding routing loops, even when links break and a node is no longer reachable to supply its own information about its sequence number. A destination becomes unreachable when a link breaks or is deactivated. When these conditions occur, the node detecting the condition increments the destination's sequence number and the metric in the route table entry is assigned to be infinite.

3.2 MAODDP Terminology

Active route: A routing table entry with a finite metric in the Hop Count field. A routing table may contain entries that are not active (invalid routes or entries). They have an infinite metric in the Hop Count field. Only active entries can be used to forward data packets. Invalid entries are eventually deleted.

Broadcast: Broadcasting means transmitting to the IP Limited Broadcast address. A broadcast packet may not be blindly forwarded, but broadcasting is useful to enable flooding.

Forwarding node: A node that agrees to forward packets destined for another destination node, by re-transmitting them to a next hop which is closer to the uni-cast destination along a path which has been set up using routing control messages.

Forward route: A route set up to send data packets from a source to a destination.

Source node: A node that initiates an MAODDP message that is the processed and possibly retransmitted by other nodes in the ad-hoc networks. For instance, the node initiating a RREQD and flooding the RREQD message is called the source node of the RREQD message.

Reverse route: A route set up to forward an acknowledged (ACK) packet back to the source from the destination or from an intermediate node having a route to the destination.

Flood: Flooding means to send a message to every node of the ad-hoc networks or to every node in an region of the ad-hoc network. In MAODDP, a message is flooded by iterated use of broadcast, for which receivers must also rebroadcast after their processing steps have been completed for that message.

3.3 MAODDP Operations

Route request and data delivery packet (RREQD), acknowledged message (ACK) and Route Error (RERR) messages for uni-cast communication towards a destination, and how the message data are handled. In order to process the messages correctly, certain state information has to be maintained for the route table entries for the destinations of interest.

3.3.1 Maintaining Sequence Numbers

MAODDP depends on each node in the network to own and maintain a sequence number to guarantee the loop-freedom of all the routes towards that node. A node increments its own sequence number in two circumstances:

- Immediately before a node originates a RREQD flood, it MUST increment its own sequence number. This prevents problems with deleted reverse routes to the source of a RREQD.

- Immediately before a destination node broadcast an acknowledged (ACK) message in response to a route request and data delivery packet (RRQED), it MUST update its own sequence number to the maximum of its current sequence number and the destination sequence number in the acknowledged packet (ACK).

Every route table entry at every node MUST include the latest information available about the sequence number for the IP address of the destination node for which the route table entry is maintained. This sequence number is called the "destination sequence number". It is updated whenever a node receives new information about the sequence number from RREQD data delivery packet, acknowledged packet, or RERR messages that may be received related to that destination.

The only other circumstance in which a node may change the destination sequence number in one of its route table entries is in response to a broken or expired link to the next hop towards that destination. The node can easily determine which destinations use a broken next hop by consulting its precursor lists for the next hop. In this case, for each destination which uses the next hop, the node increments the sequence number and puts the Hop Count to be "infinity"

In summary, a node may change the sequence number for a particular destination only if:

- It is itself the destination node, and offers a new route to itself
- It receives an MAODDP messages i.e. RREQD OR ACK, with new information about the sequence number for some other destination node
- The path towards the destination node expires or breaks.

3.3.2 Maintaining Route Table Entries and Route Utilization Records

For each valid route maintained by a node (containing a finite Hop Count metric) as a routing table entry, the node also maintains a list of precursors that may be forwarding packets on this route. These precursors will receive notifications from the node in the event of detection of the loss of the next hop link. The list of precursors in a routing table entry contains those neighboring nodes to which a route reply was generated or forwarded.

When a node receives an MAODDP control packets from a neighbor, it checks its route table for an entry for that neighbor. In the event that there is no corresponding entry for that neighbor, an entry is created. The sequence number is either determined from the information contained in the RREQD, or else it is initialized to zero if the sequence number for that node can not be determined. The lifetime for the routing table entry is either determined from the RREQD, or it is initialized to MY_ROUTE_TIMEOUT. The hop count to the neighbor is set to one.

Each time a route is used to forward a data packet, its Lifetime field is updated to be no less than the current time plus ACTIVE_ROUTE_TIMEOUT. Since the route between each source and destination pair are expected to be symmetric, the Lifetime for the previous hop, along the reverse path back to the IP source, is also updated to be no less than the current time plus ACTIVE_ROUTE_TIMEOUT.

3.3.3 Generating Route Requests

A node floods a RREQD when it determines that it needs a route to a destination and does not have one available. This can happen if the destination is previously unknown to the node, or if a previously valid route to the destination expires or is broken (i.e., an infinite metric is associated with the route). The Destination Sequence Number field in the RREQD is the last known destination sequence number for this destination and is copied from the Destination Sequence Number field in the routing table. If no sequence number is known, a sequence number of zero is used. The Source Sequence Number in the RREQD is the node's own sequence number. The Flooding ID field is incremented by one from the last Flooding ID used by the current node. Each node maintains only one Flooding ID. The Hop Count field is set to zero.

Before flooding the RREQD, the source node buffers the Flooding ID and the Source IP address of the RREQD for FLOOD_RECORD_TIME milliseconds. In this way, when the node receives the packet again as it is flooded by its neighbours, it will not reprocess and re-forward the packet.

A source node often expects to have bi-directional communications with a destination node. In such cases, it is not sufficient for the source node to have a route to the destination node; the destination must also have a route back to the source node. In order for this to happen as efficiently as possible, any generation of an acknowledge packet by the destination node for delivery to the source node, should be accompanied by some action which notifies the destination about a route back to the source node. The source node selects this mode of operation in the intermediate nodes by setting the `G' flag. After broadcasting a RREQD, a node waits for a acknowledged packet (ACK). If the acknowledged packet is not received within NET_TRAVERSAL_TIME milliseconds, the node MAY try again to flood the RREQD, up to a maximum of RREQD_RETRIES times. Each new attempt MUST increment the Flooding ID field.

Data packets waiting for a route (i.e., waiting for a acknowledged packet after route discovery and data delivery packet has been sent) SHOULD be buffered. The buffering SHOULD be FIFO. If a (RREQD) has been flooded RREQD_RETRIES times without receiving any ACK, all data packets destined for the corresponding destination SHOULD be dropped from the buffer and a 'Destination Unreachable' message delivered to the application.

3.3.4 Processing and Forwarding RREQD

When a node receives a flooded RREQD, it first checks to determine whether it has received a RREQD with the same Source IP Address and Flooding ID within at least the last FLOOD_RECORD_TIME milliseconds. If such a RREQD has been received, the node silently discards the newly received RREQD. The node always creates or updates a reverse route to the Source IP Address in its routing table. If a route to the Source IP Address already exists, it is updated only if either

- The Source Sequence Number in the RREQD is higher than the destination sequence number of the Source IP Address in the route table, or
- The sequence numbers are equal, but the hop count as specified by the RREQD, plus one, is now smaller than the existing hop count in the routing table.
- This reverse route would be needed in case the node receives an eventual ACK back to the node which originated the RREQD (identified by the Source IP Address). When the reverse route is created or updated, the following actions are carried out:
- The Source Sequence Number from the RREQD is copied to the corresponding destination sequence number.
- The next hop in the routing table becomes the node transmitting the RREQD (it is obtained from the source IP address in the IP header and is often not equal to the Source IP Address field in the RREQD message).
- The hop count is copied from the Hop Count in the RREQD message and incremented by one.

3.3.5 Generating Acknowledge (ACK) by the Destination

If a RREQD is successfully deliver to the intended destination. It is the responsibility of the destination node to issue acknowledge packet (ACK) back to the source node. Destination node MUST update its own sequence number to the maximum of its current sequence number and the destination sequence number in the RREQD packet. The destination node places the value zero in the Hop Count field of the ACK message.

The destination node copies the value MY_ROUTE_TIMEOUT into the Lifetime field of the ACK. Each node MAY reconfigure its value for MY_ROUTE_TIMEOUT.

3.3.6 Forwarding Acknowledge packet

When a node receives an ACK message, it first increments the hop count value in the ACK by one, to account for the new hop through the intermediate node. It then compares the Destination Sequence Number in the message with its own copy of destination sequence number for the Destination IP Address in the ACK message. The forward route for this destination is created or updated only if:

- The Destination Sequence Number in the ACK is greater than the node's copy of the destination sequence number, or (ii) the sequence numbers are the same, but the route is no longer active or the incremented Hop Count in ACK is smaller than the hop count in route table entry. If a new route is created or the old route is updated,
- The next hop is the node from which the ACK is received, which is indicated by the source IP address field in the IP header; the hop count is the Hop Count in the ACK message plus one; the expiry time is the current time plus the Lifetime in the ACK message; the destination sequence number is the Destination Sequence Number in the ACK message.

3.3.7 Route Error Messages

A node initiates a RERR message in three situations:

1. if it detects a link break for the next hop of an active route in its routing table or
2. if it gets a data packet destined to a node for which it does not have an active route, and has already made an attempt at local repair, or
3. If it receives a RERR from a neighbor for one or more active routes.

For above given two cases, for each unreachable destination the node copies the value in the Hop Count route table field into the Last Hop Count field, and marks the Hop Count for this destination as infinity, and thus invalidates the route.

4. CONCLUSION

In this paper, we have presented our thoughts about future deployment of mobile ad hoc networks in a home environment. We have also discussed routing problem in general and have g covered a brief description of mobile ad-hoc on demand data delivery protocol as a routing solution for mobile ad hoc networks.

Unlike fixed wireless networks, mobile ad hoc networks operate without the aid of any fixed infrastructure. This approach could make it possible in near future to have networking environment in our home, business and other environment at a much reduced cost. However there are several issues which need to be resolved in order to see successful deployment of mobile ad hoc network under various environments. Our next step is to further investigate the possible use of MAODDP for mobile ad hoc networks in their deployment for future home networking environments.

REFERENCES

[1] Humayun Bakht, Madjid Merabti, and Robert Askwith. Centralized frame for routing in mobile ad-hoc networks. in International Conference on Computer Communication (ICCC2004). September, 2004. Beijing, China.

[2] Humayun Bakht, Understanding mobile ad hoc network, in Computing Unplugged. June 2004. p. 2.

[3] Humayun Bakht, Data Communication in mobile ad-hoc networks, in Computing Unplugged. September 2004. p. 2.

[4] Humayun Bakht, Group communications in mobile ad hoc networks, in Computing Unplugged. November 2004. p. 1.

[5] Humayun Bakht, Technical aspects of mobile ad-hoc networks, in Computing Unplugged. June 2004. p. 2.

[6] Humayun Bakht, A focus on the challenges of mobile ad hoc network, in Computing nplugged. August 2004. p. 2.

[7] Humayun Bakht, et al. Multicasting in mobile ad hoc networks. in 9th CDMA International Conference. 25-28 October 2004. Seoul , Korea.

[8] Humayun Bakht, Importance of secure routing in mobile ad hoc networks, in Computing Unplugged. August 2004. p. 2.

[9] Humayun Bakht, Some characteristics of mobile ad-hoc networks, in Computing Unplugged. July 2004. p. 2.

[10] Sulabh Agarwal, et al., Route-Life time Assessment Based Routing (RABR) Protocol for Mobile Ad-Hoc Networks. IEEE International Conference on Communications (ICC), June 2000.

[11] Stefano Basagni, Imrich Chlamtac, and Violet R. Syrotiuk, A Distance Routing Effect Algorithm for Mobility (DREAM). Proceedings of the ACM/IEEE International Conference on Mobile Computing and Networking (MobiCom), Dallas, TX,, October 25-30, 1998: p. 76-84.

[12] Tsu-Wei Chen and Mario Gerla, Global State Routing: A New Routing Scheme for Ad-hoc Wireless Networks. In Proceedings of IEEE ICC '98, 1998.

[13] M.S. Corson and A. Ephremides, Lightweight Mobile Routing protocol (LMR)A distributed routing algorithm for mobile wireless networks. Wireless Networks, 1995. 1.

[14] Benjie Chen, et al., Span: An Energy-Efficient Coordination Algorithm for Topology Maintenance in Ad Hoc Wireless Networks. the 7th ACM International Conference on Mobile Computing and Networking, Rome, Italy, July 2001.
[15] Nitin Vaidya., Tutorial on Mobile Ad Hoc Networks: Routing, MAC and Transport Issues. 2001 MobiCom (Rome) and 2000 MobiCom (Boston), 2000.
[16] Humayun Bakht, Routing protocols for mobile ad hoc networks, in Computing Unplugged. July 2004. p. 2.
[17] Humayun Bakht, Bluetooth a commercial application of mobile ad hoc networks, in Computing Unplugged. November 2004. p. 2.
[18] Humayun Bakht, Future of mobile ad hoc networks, in Computing Unplugged. October 2004. p. 1.
[19] Humayun Bakht, Sensor networks and ad hoc networking, in Computing Unplugged. Oct 2004. p. 1.
[20] Humayun Bakht, Some applications of mobile ad-hoc network, in ComputingUnplugged. September 2004. p. 1.

ASSISTIVE ENVIRONMENTS

Integration of heterogenous networks to support Ambient Intelligence

J. Abascal, J. L. Sevillano, A. Civit, G. Jiménez, J. Falcó

J Abascal, University of the Basque Country-EHU. Donostia-San Sebastián, Spain. J.L Seviallano, A Civit and G Jimenez, ETSI Informática. Universidad de Sevilla. Spain. J Falco, CPS, Universidad de Zaragoza, Spain

Abstract: In the last decade, Assistive Technology researchers have studied and developed accessible and adaptable environments intended to enhance the autonomy and the quality of life of the people with disabilities and elderly people. Their efforts, technologically rooted in the Ubiquitous Computing technologies, are firmly supported by the Ambient Intelligence concept. In the assistive scenario the main problem that researchers have to face is the integration of diverse local and remote services that were previously provided through different network technologies. In this paper we identify the main issues that may influence design decisions regarding the integration of -and the interaction among- different subsystems in the context of assistive technologies.

Key words: Assistive technology, networks, ambient intelligence.

1. INTRODUCTION

The concept of Ambient Intelligence (AmI) defines a set of both reactive and proactive properties of an environment in order to enable individuals and devices to easily interact among them [1]. Ducatel et al [2] foresee some scenarios for AmI where Ubiquitous access, Context awareness, Natural interactions, and also the need to support heterogeneous systems are particularly important [3].

AmI objectives are totally coincident with the aims of Assistive Technology to produce accessible smart environments in order to enhance the autonomy and the quality of life of people with physical, sensorial or cognitive restrictions due to diverse causes, such as disability or aging. In this way, many services have been developed to assist people in their diverse needs. However, a single person frequently uses diverse services delivered by means of different networks and devices. A smart environment should integrate and provide all the services by means of a single device.

2. INCLUDING SCENARIOS FOR PEOPLE WITH SPECIAL NEEDS

Therefore, one of the main challenges of AmI in the context of assistive technologies is the integration of many pre-existent heterogeneous digital devices and services and their networks into everyday environments. The whole system should be able not only to support the interaction of heterogeneous networks, but also the integration of diverse services and applications. The reason is that assistive technologies are very varied when attending needs due to individual diversity and to temporal variations. In addition, the diverse devices needed by a single person are usually designed by different manufacturers using different technologies and devoted to varied applications. Individual diversity and variability increase heterogeneity in subsystem development, both in terms of applications and services, in a kind of vicious circle.

We will look at this problem later on, but first let's illustrate through a typical scenario how the AmI concept will influence assistive technologies:

> A wheelchair user with low vision restrictions and wearing a mobile computer enters an unknown building (let's say the City Hall), provided with Ambient intelligent facilities. As soon as he/she gets into the building, the Ambient Intelligence System (AmIS) discovers his/her presence and announces the services that can be used. Additionally, it investigates what services the user's system can provide to the community. Both systems also exchange information related to preferences, special needs (cognitive, sensorial, physical and communication abilities), and technological constraints (display resolution, voice, text, pixel-based, bandwidth, computing power, etc.).
>
> The AmIS offers communication with a remote information centre that appears in the display of the user, adapted to its physical and cognitive characteristics (text menu, voice, icons…). It also offers a video guided tour that our user rejects. Our user asks the information centre where he/she must go to see what had happened with his/her water bill. He/she

chooses the voice and video interface with the person in the information office. After knowing where to go it appears on its display the possibility of a route-guiding tool, which the user accepts. It uses a location service and gives information via text messages. The user is located and receives his/her position, together with a message of where to go now. The AmIS has calculated the best path to follow, taking into account the user constraints and the information of occupation in the building at that time of day. When he/she enters an elevator communication with the system is lost. The AmIS uses elevator information that sends the vertical position. Originally this feature was only designed and used by the technical personnel of the elevator company, but now it can be used to provide both the user and the route guiding system with the information of the floor he/she is at. The elevator doors also adapts to the especially long timing of our user, for it receives the information of this special characteristic.

There is an especially narrow door he/she must cross, so the user accepts the service of semiautomatic wheelchair navigation he/she was offered at the entrance, which helps him/her with that specific manoeuvre.

Finally he/she arrives at the destination and solves the water fee problem by direct personal interaction. The user's mood is better now that his/her worries about the water fee are solved, so he/she accepts some music service while leaving the building in the same way he/she entered.

3. DESIGN ISSUES

Evidently, assistive technologies present specific characteristics regarding the integration of different and heterogeneous subsystems. Note that AmISs not only need the inclusion of different existing subsystems but should also allow their interaction so that each subsystem (offering a particular service) can take advantage of the services that are included in the environment. For instance, wheelchair users who want to access assistive services through personalized interfaces from different locations, even in unfamiliar environments; services for monitoring users' location in terms of safety or support in hospitals or residences; services to provide information adapted to user needs or abilities in museums; etc. We identify three main issues with regard to the integration of -and interaction among- different subsystems in this context:

Mobility: There are several aspects regarding this issue. First, users should be able to interact with services and sub-systems at anytime, from any place, and using a personal (adapted) interface carried by the user. These mobile devices used to have scarce resources: bandwidth, computational and battery power, screen size, etc. However, advancements in this field are enormous and there are now commercially available handheld devices (such as PDAs, mobile telephones and portable PCs) of acceptable performance at relatively low cost. An additional problem is to provide communications across heterogeneous and dynamically changing links and networks. Wireless personal (e.g. Bluetooth) and local (e.g. Wi-Fi) area networks now permit low-cost commercial solutions for this type of communication, but there are still open problems like efficient roaming, reachability, intermittent failures, fault tolerance, etc., in order to achieve seamless access. Finally, it is not just the problem of communicating subsystems. Further issues are how to share services and information among heterogeneous sub-systems "on the move".

Heterogeneity: As we discussed above, this is a major problem in this field. In fact, the current situation is characterized by what could be called "islands of functionality": solutions adapted to specific users in particular environments (such as home or collective residences). With regard to hardware devices and wireless/wired links, standardization is the norm. However, this is not the case with assistive devices and applications. In order to successfully integrate these heterogeneous subsystems, generic descriptions of devices, services and/or context (location, personal, etc.) are needed in a way that is independent from the applications. In assistive technologies, one of the main issues is a generic and universal description of personal abilities and characteristics. The latter also has to do with how adapted assistive services are offered to the right persons. Excess or useless content must be avoided. For instance, audio environmental descriptions may only be required by visually impaired people.

Context awareness: Mobility and environment and personal variability in these systems require a context-dependent computation, that is, computation that uses information to characterize the situation (location, time, personal) of an entity (person, place, object) [4]. This is not a constraint as such. On the contrary, it may result in one of the added values due to this interaction at all levels. From the point of view of the integration of context-aware subsystems, a common context representation is required, and also a common way to interchange this context information among subsystems.

4. DESIGN DECISIONS FOR SUB-SYSTEM INTERACTIONS

In a previous paper [3], we proposed a layered architecture for the development and integration of heterogeneous sub-systems into Ambient Intelligent Systems (AmISs). Since the systems to be connected are also distributed, we identified three levels that help to model the interaction among systems with different types of interactions depending on the level.

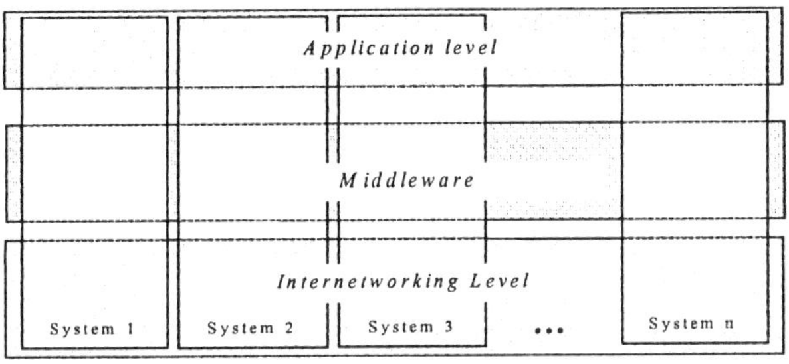

Figure 1. Interaction Scheme

The interactions are named *Interconnectivity*, *Interoperability* and *Interfunctionality* at the Internetworking, Middleware and Application levels, respectively. In the following sub-sections, we discuss our main design decisions at every level, taking into account the identified issues described in section 2.

4.1 Interconnectivity

We define *interconnectivity* as the ability to interact at the internetworking level, including all the lower level functions (approximately equivalent to the transport, network, data link and physical OSI layers [5]). At this level of interactions, the necessary mobility implies that communications should be *asynchronous*. In other words, interconnectivity cannot be guaranteed at any time. Mobility requires wireless connections that may suffer frequent connection losses, what forces asynchronous communications. Clients asking for a service and devices offering it may not be connected at the same time. The communication paradigm should be *connectionless* (vs. connection oriented), well suited for intermittent connections.

On the other hand, heterogeneity makes a *Nomadic* system better that an *Ad-Hoc* system: a backbone fixed infrastructure plus a number of mobile devices connected through wireless links is a better solution. A robust and contrasted solution is a backbone network based on the IP protocol, which has demonstrated its success in the interconnection of heterogeneous devices (a good example is the Internet). Most devices can be connected through this IP network while secondary, maybe simpler, devices (e.g. sensors) may be connected using non-IP communications. In this case a gateway is used to interconnect IP and non-IP sub-networks. This configuration allows both sensing and collecting context information (context awareness). Furthermore, it permits environmental control in a remote mode via a web page as well as direct Internet access in *home automation* through *Residential Gateways*. It may seem that the required mobility would make ad-hoc communications more appropriate. However, we think that a nomadic system presents several advantages. It simplifies connection establishment and roaming. Mobile devices usually have limited resources (computing power, bandwidth, memory, etc) so some of the complexity needed or desired for some functions can be placed in fixed systems. The infrastructure can maintain knowledge about device characteristics and manage coherent device interactions.

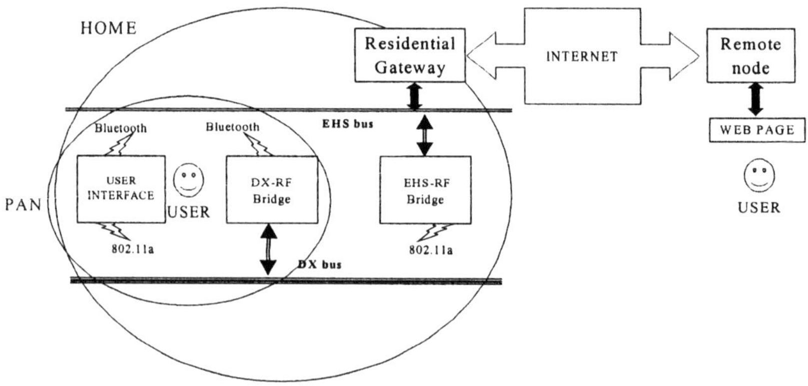

Figure 2. Networks interconnection in the DomoSilla system

As an example, figure 2 shows the set of bridges that provide the interconnection level in the *DomoSilla*[21] project. The user interacts with the AmIS through a mobile user interface handled with the wheelchair input system (usually a kind of joystick). Through this interface, the user controls both the wheelchair and the domotic system. The network infrastructure

[21] *DomoSilla*, "Study, evaluation and design of an interconnection system between local network for wheelchairs control (DXBus) and domotic network (EHS)", funded by the Spanish Ministry of Science and Technology under the grant No. TIC2000-0087-P4.

includes two wireless networks. First, there is a Bluetooth personal area network (PAN) to communicate the user interface with the computer in the wheelchair through a bridge between the radiofrequency network and the real time bus (DX) that controls the wheelchair. Second, there is a Wi-Fi home area network to interconnect the domotic system and the wheelchair. For this purpose another bridge between the radiofrequency network and the domotic EHS bus (a Powerline-like bus used for the control of home devices) is available. In addition, it is also possible to remotely interact with domotic system through a web page. For this purpose there is a residential gateway that connects the home EHS bus to the Internet.

4.2 Interoperability

The concept of interoperability is widely used to describe interaction among devices at all levels [6], including control, configuration and information sharing in different formats [7]. To distinguish this term from *interconnectivity*, we consider *interoperability* related to the sharing services at the middleware level, such as import/export services [8]. Interoperability can provide a set of services to all elements. In the literature there is a general agreement about the functions that should be related to interoperability [5, 9], including dynamic service discovering (periodically or triggered by determined events [10]), service description (including actions that may be performed, properties that may be useful, even devices for which connection was not planned), and service control (actions and modifications of state or attributes of a service in a sub-network from another device connected to a different sub-network).

A number of available architectures can support these functions, and we will now revise some of them taking into account the specific characteristics of the environments and applications for Ambient Intelligence. HAVi[22] specification allows registration of new devices as they add to the network, so the rest of devices know what new functions they may perform. Most common A/V (Audio/Video) functions have some standard APIs (Application Programming Interfaces) to enable other devices to use them. The system allows the installation of applications and user interface software in each device in an automatic way. These APIs are specified in a generic C-like language named IDL (*Interface Definition Language*). Specification is independent of language although they are usually implemented in Java [11]. A disadvantage of HAVi is that it assumes the communication channel is IEEE1394, a wired connection with enough bandwidth to transmit audio and

[22] http://www.havi.org/

video, and so its architecture may show limitations when used in a wide range of applications, especially nomadic systems [12].

Another *middleware* described in literature is Jini[23]. Jini gives support for all the described structure concerning service discovery and sharing of services by different clients. Mascolo et al. point to its heavy dependency on Java and its assumption of the existence of a fixed infrastructure as its main drawbacks [5]. On the other hand, some authors think that Jini is not especially adequate for limited computational resources devices and, mainly, for mobile devices [12,13]. Nevertheless, the recent release of very small Java virtual machines opens the opportunity to the inclusion of very small embedded devices that previously could not be integrated.

On the contrary, there are many devices on the market based on UPnP (Universal Plug and Play)[24], including all Windows XP-based systems. UPnP supports all mentioned functions including the dynamic connection of a device to a network, services offering and discovery, everything based on a unified description of functions and attributes of services through XML (*eXtended Mark-up Language*) documents. There are two factors that make UPnP especially attractive from our point of view: one is the use of IP protocols at the lowest level; second is the use of open and standard protocols.

As IP is used for the lowest level, the first phase of interaction among UPnP systems is *Addressing*. This is the mechanism that devices have to obtain an address that makes them visible in all the system. At first, all UPnP devices incorporate a DHCP (*Dynamic Host Configuration Protocol*) client to obtain an IP address. If there is no answer from a DHCP server, it chooses an address and checks it wasn't used through ARP (*Address Resolution Protocol*). In any case, once the device has an IP address, it can proceed with following phases related to discovery functions and service control, that are performed using an also open and simple protocol named SSDP (*Simple Service Discovery Protocol*).

IP based communications are made using HTTP protocol or any of its versions. HTTP is message based (asynchronous communication), which is especially adequate for nomadic systems, that may suffer unpredictable disconnections. Particularly, UPnP considers the usage of HTTPU (or HTTPMU for *multicast* messages, very useful when offering services to the whole system), that is HTTP version over UDP. UDP gives functionality equivalent to the transport (TCP) but is simpler, and, most importantly, non-connection oriented. So, the use of the set HTTPU / UDP / IP is, in our

[23] http://www.jini.org

[24] http://www.upnp.org

opinion, an adequate solution for asynchronous communication in nomadic systems.

Development of UPnP devices is greatly facilitated thanks to tools like *Intel Service Author, Device Builder and Device Validator* [14]. Siemens also has some tools, and there are even some freeware developments for Linux. A disadvantage of UPnP may be the difficulty for simple devices to interpret XML descriptions of devices and services. Moreover, these devices may find difficulties in supporting all the pile of UPnP protocol. A feasible solution for simple devices and/or slow connections is the use of SCP. This is a no TCP/IP based protocol, but it uses the same schemes and models as UPnP, allowing easy interoperation among UPnP and SCP devices via simple bridges.

Finally, it may be worthwhile to name other interoperability solutions, not considered here due to their limitation to a set of applications, such as *Salutation* in office environments, or to be limited to specific functions or technologies, as SDP (*Service Discovery Protocol)* of Bluetooth.

As there are many available technologies at *middleware* level, and they are not compatible with each other, the heterogeneity problem remains as a major difficulty for AmISs. Already in the literature there are descriptions of efforts to overcome heterogeneity at *middleware* level, such as OSGi (Open Services Gateway specification) initiative [15]. It gives a common structure at application level that is independent of the middleware technology used. Actually what OSGi offers is a specification for a somehow centralized gateway that allows the use of services of different technologies, as Jini and UPnP. Something similar has been proposed using techniques already used in the Internet, through the concept of *Virtual Overlay Networks* [12]. Generally speaking, this idea of a gateway is placed over the level of middleware, closer to the application level, so it differentiates somehow from the former interoperability concept.

Whatever solution is chosen, several design considerations can be made according to the previous sections. If nomadic systems are used at the interconnectivity level, then the fixed infrastructure may support some interoperability functions. For instance, a central unit (*Residential Gateway*) may gather and distribute some services. Also, the context captured by fixed elements (for instance the closest service access point/provider in public environments like airports) can approximate that of the mobile client. Mobile devices could therefore obtain an IP address, connect to a fixed system and collect/send services and context information in order to interoperate. This interoperation also requires abstract context representation as well as generic, common protocols, including common formats for multimedia content distribution.

On the other hand, a fully distributed solution based on an ad-hoc network would be still possible, with devices advertising services (multicast) and collecting context information, but it would require more powerful devices and complex protocols.

4.3 Interfunctionality

Interaction among subsystems at the interoperability level allows the services to be discovered and shared as a *syntactic* interaction, without considering their "meaning". We propose a higher level of *semantic* interactions, similar to those described in the literature for other fields [16, 17, 18]. This need is emphasized by the ISTAG as being one of the key issues in future AmISs: "...[the] need for horizontal integration that goes far beyond the simple dynamic service discovery mechanisms" [1]. Obviously the usefulness of the service and its applicability should be described in more detail. Several languages have been proposed, such as RDF (Resource Description Language) [19] and future work will address the issue of which language best suits AmI applications.

Interfunctionality would add two main values for subsystem interactions. First, the semantic descriptions allow us to pre-select the services previewed as useful for the applications of a particular subsystem. This allows a selection for limited resource subsystems when entering environments with higher richness of services or high complexity. Thus, the system can have a set of available services, and every subsystem can choose the most useful or manageable through semantic interaction. Although users may be active (selecting and configuring the most useful and appropriate services), pre-selection is often desirable to offer adapted assistive services to the right people. Excess or useless contents must be avoided because the users will reject them [20].

Second, *interfunctionality* can be used to adapt or empower the functionality of existing applications according to new services. The new applications may become available based on the new services. As an example, let us consider a wheelchair user with an assisted navigation application who enters a building with a positioning and location system. Under user supervision, the navigation system can ask for information about the path to follow to reach the desired destination, and then incorporate the positioning service to follow it. The functionality of navigation assistance has improved greatly due to the new services and involve more than just avoiding bumping into objects and helping when passing doors.

Interfunctionality among heterogeneous systems presents new challenges. First, there is the question of mixing services from incompatible technologies. Common descriptions and the use of a kind of "virtual services" may be needed. Also, compatible authentication and authorization mechanisms should be used. However, at this level of interactions, probably the main difficulty is the use of new functions with little or no advance planning. It may be difficult to foresee potential uses of services that were not considered at the designing phase. Further, new services should offer different complexity descriptions and/or interfaces both attending user abilities (cognitive and sensorial impairments) and technological constraints (display size and resolution, bandwidth, voice, text...). Also, not all services of environmental control may be made available to the user interface due to technological constraints, user abilities or privileges. Services and devices description, together with appropriate context characterization, should allow this pre-selection to be performed automatically. On the other hand, the internal structure of applications should be modular and parametrical to allow interaction and modification of their functionality.

Four laboratories from four different Spanish universities (U. of the Basque Country, U. of Seville, U. of Saragossa and Polytechnic U. of Barcelona) are engaged in the development of the *interfunctionality* level of a complex AmIS, through Heterorred Project. Parts of this system are being tested in a residential home for elderly dependent people.

5. CONCLUSIONS

The design of smart environments for people with disabilities and elderly people is firmly supported by the Ambient Intelligence concept. This particular scenario presents special difficulties due to the need to integrate new and pre-existing heterogeneous services and devices, created to assist people with physical perceptive or cognitive restrictions. The main technological problem is to integrate heterogeneous wired and wireless networks that support diverse services. In this paper the three levels of integration - the physical network, middleware, and service - are identified and analyzed. For each level a model of interaction is defined: interconnection, interoperability and *interfunctionality*. Available implementations of the first two levels (interconnection and interoperability), both academic and commercial, have been discussed. The concept of *interfunctionality* has been also introduced.

ACKNOWLEDGEMENTS

The research presented in this paper has been developed by the Universities of Seville, Saragossa and Basque Country within the project *Heterorred*

"Study and development of a heterogeneous personal area network for interoperability and access to wireless services and communications", funded by the Spanish Ministry of Science and Technology under grant No. TIC2001-1868-C03.

REFERENCES

[1] ISTAG Draft Consolidated Report: Ambient Intelligence: from vision to reality. Sept. 2003. http://www.cordis.lu/ist/istag.htm.

[2] Ducatel K., Bogdanowicz M., Scapolo F., LeijtenJ., Burgelman J-C. ISTAG Scenarios for Ambient Intelligence in 2010. February 2001. IPTS-Seville. ftp://ftp.cordis.lu/pub/ist/docs/ istagscenarios2010.pdf

[3] Sevillano J. L., Falcó J., Abascal J. , Civit-Balcells A. , Jiménez G. , Vicente S. , Casas R.: On the Design of Ambient Intelligent Systems in the Context of Assistive Technologies: ICCHP 2004. 9th International Conference on Computers Helping People with Special Needs, Paris 2004. LNCS. Springer 2004.

[4] A.K. Dey: Understanding and Using Context. Personal and Ubiquitous Computing 5(2001)20-24.

[5] Mascolo C., Capra L., Emmerich W.: Mobile Computing Middleware. Tutorial in IEEE MWCN'02. Stockholm, Sweden. Sep. 2002.

[6] Carpenter B. E.: Interoperability among Heterogeneous Communications Networks- an IETF Perspective. Computer Standards & Interfaces 20. 1998. Pp. 147-149

[7] Rasheed Y., Edwards J., Tai C.: Home Interoperability Framework for the Digital Home. Intel Technology Journal 6 (2002) 5-16.

[8] Cho S.Y.: Framework for the Composition and Interoperation of the Home Appliances based on Heterogeneous Middleware in Residential Networks. IEEE Trans. Consumer Electronics 48-3. (2002) 484-489.

[9] Miller B. A., Nixon T., Tai C., Wood M.D.: Home Networking with Universal Plug and Play. IEEE Communications Magazine 39 (2001) 104-109.

[10] Wils A. et al.: Device Discovery via Residential Gateways. IEEE Trans. Consumer Electronics, 48-3(2002)478-483

[11] Lea R., Gibbs S., Dara-Abrams A., Eytchison E.: Networking Home Entertainment Devices with HAVi. IEEE Computer 33 (2000) 35-43.

[12] Nakajima T. et al.: A Virtual Overlay Network for Integrating Home Appliances. Proc. 2002 Symp. Applications and the Internet. Nara, Japan. Jan. 2002. Pp. 246-253.

[13] Helal S.: Standards for Service Discovery and Delivery. IEEE Pervasive Computing 1. (2002) 95-100.

[14] http://www.intel.com/technology/upnp/index.htm.

[15] Dobrev P., Famolari D., Kurzke C., Miller B. A.: Device and Service Discovery in Home Networks with OSGi. IEEE Communications Magazine 40 (2002) 86-92.

[16] Stephanidis C., Savidis A.: Universal Access in the Information Society: Methods, Tools, and Interaction Technologies. Universal Access in the Information Society. June, Vol.1. 2001. Pp. 40-55.

[17] Paolucci, M.; Sycara, K.: Autonomous Semantic Web services. IEEE Internet Computing, 7-5(2003) 34 -41.

[18] Special issue on "Web Services". Computer, vol.36, Oct. 2003.

[19] S. Decker, P. Mitra, S. Melnik: Framework for the Semantic Web: An RDF Tutorial. IEEE Internet Computing 4-3(2000) 68-73.
[20] E. Kaasinen: User Needs for Location-aware Mobile Services. Personal and Ubiquitous Computing 7:(2003)70-79.